T0259723

SpringerBriefs in Applied Sciences and Technology

Forensic and Medical Bioinformatics

Series editors

Amit Kumar, Hyderabad, India
Allam Appa Rao, Hyderabad, India

More information about this series at http://www.springer.com/series/11910

P. Venkata Krishna · Sasikumar Gurumoorthy
Mohammad S. Obaidat

Internet of Things
and Personalized Healthcare
Systems

 Springer

P. Venkata Krishna
Department of Computer Science
Sri Padmavati Mahila Visvavidyalayam
Tirupati, Andhra Pradesh, India

Mohammad S. Obaidat
Department of Computer
and Information Science
Fordham University
Bronx, NY, USA

Sasikumar Gurumoorthy
Department of Computer Science
and Systems Engineering
Sree Vidyanikethan Engineering College
Tirupati, Andhra Pradesh, India

ISSN 2191-530X   ISSN 2191-5318 (electronic)
SpringerBriefs in Applied Sciences and Technology
ISSN 2196-8845   ISSN 2196-8853 (electronic)
SpringerBriefs in Forensic and Medical Bioinformatics
ISBN 978-981-13-0865-9   ISBN 978-981-13-0866-6 (eBook)
https://doi.org/10.1007/978-981-13-0866-6

Library of Congress Control Number: 2018957056

This Springer imprint is published by the registered company Springer Nature Singapore Pte Ltd.
The registered company address is: 152 Beach Road, #21-01/04 Gateway East, Singapore 189721, Singapore

Contents

Chapter 1
Sensitivity Analysis of Micro-Mass Optical MEMS Sensor for Biomedical IoT Devices

Mala Serene, Rajasekhara Babu and Zachariah C. Alex

Abstract Micro-electromechanical systems (MEMS) have tremendous applications in the field of biomedical and chemical sensors. There are different readout techniques like piezo-resistive and piezo-electric which are used to measure the stimuli absorbed by the cantilever into electrical signals. In this paper, we used the open-source Ptolemy software to model MOEMS sensor with novel optical read out. To enhance the deflection and sensitivity, four micro-mass optical MEMS sensor models were developed using four different shapes of the cantilever. The detectable mass range measured by the triangular cantilever using parylene as material is 50.97 µg–23.996 mg.

1.1 Introduction

Microcantilever sensors offer a highly promising area to sense various physical stimuli, chemical vapors, and measure very small masses. The different shapes of the cantilever are used to detect different diseases, chemical, and micro-masses. Many readout methods like piezo-resistive and piezo-electric are used to convert the stimuli present on the cantilever. The optical lever method uses the atomic force microscopy which can accurately measure the deflection of the cantilever when compared to the above-mentioned techniques [1].

Many researchers reported that the mostly used is rectangular-shaped microcantilever for sensing applications [2]. Ansari et al. [3] made a study on rectangular and trapezoidal shapes of cantilever and measure the deflection of the cantilever, fundamental resonant frequency, and maximum stress. Profitable FEM ANSYS software is used to analyze the designs. Three models in each shape have been analyzed, and sensitivity was measured. The paddled trapezoidal cantilever has better sensitivity than others. Hawari et al. [4] studied and made a different analytical model of cantilevers to measure the stress and highest deflection fundamental resonant frequency using ANSYS software.

The mathematical model of the tapered cantilever beam was derived using the Euler–Lagrange method. For the first three modes of cantilever, the vibration

© The Author(s), under exclusive license to Springer Nature Singapore Pte Ltd. 2019 1
P. V. Krishna et al., *Internet of Things and Personalized Healthcare Systems*,
SpringerBriefs in Forensic and Medical Bioinformatics,
https://doi.org/10.1007/978-981-13-0866-6_1

amplitude and taper ratio are obtained on the nonlinear natural frequencies and presented in non-dimensional form [5]. A numerical study was done to evaluate the impact of microcantilever geometry on mass sensitivity. The mass of biological agents in liquids can be measured with different shapes of the cantilever. Using ANSYS software, the modal analysis was done on various shapes like rectangle, trapezoid, and triangle. The results indicated that resonant frequency shift is ruled by the tiny mass end of the cantilever and the width of the cantilever at the fixed end. Among the three shapes of cantilever, triangular cantilever shows the increase in mass sensitivity [6]. The T-shaped cantilever was designed and fabricated. The microcantilever sensor was actuated thin film of zinc oxide film. With the help of Rayleigh–Ritz method, the basic frequency formula for cantilever was derived and validated by simulation results. The fundamental resonant frequency and sensitivity are higher for T-shaped cantilever than rectangular cantilever [7]. A cantilever biosensor which can sense analytes in low concentrations was modeled and simulated. A new cantilever was designed and sense analytes in extreme at low concentrations [8].

A rectangular MEMS cantilever provides very less deflection, and sensitivity is very low. To improve the sensitivity and deflection, three different shapes are proposed like trapezoidal, trapezoidal beam with square step, and length-wise symmetrical tree-type microcantilever. The new design exhibits twice the deflection and higher sensitivity than conventional rectangular beam [9]. The cantilever-based biosensor was designed and simulated using PZR module. A new proposed model has a small strip near the fixed end of the SU-8 polymer cantilever that was simulated. The new model gives 2.5 times higher deflection than rectangular cantilever model, and sensitivity was improved [10]. The microcantilever was designed and simulated using finite element software COMSOL. Three different shapes like triangle, PI shape, and T shape are modeled and simulated. Triangular-shaped cantilever exhibits more sensitivity than other shapes [11]. The maximum resonant frequency was observed by the V-shaped cantilever. The cantilever with narrow strip at fixed end and wide free end gives more deflection in static mode than rectangular cantilever [12]. In this paper, micro-mass optical MEMS sensor is developed using rectangular shape as cantilever beam. The fundamental frequency of the cantilever beam and its sensitivity are measured. To improve the sensitivity of the cantilever, various parameters like length, thickness, shape, and material of the cantilever can be changed. In this paper, shape of the cantilever is modified to improve of sensitivity, and three optical micro-mass optical MEMS sensor mathematical models are developed and simulated using three different shapes of the cantilever using the open-source framework Ptolemy.

1.2 Modeling and Simulation

A. Modeling and simulation play a key role in the silicon electronic industry. Through the modeling process, the system can developed and its behavior can be studied in a particular environment [13]. As it is already mentioned that the system-level software is not available for optical MEMS, Ptolemy II is an open-source framework chosen

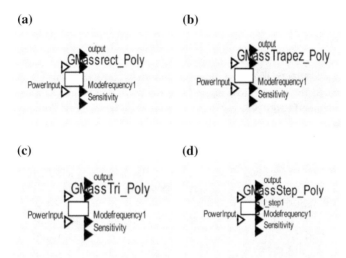

Fig. 1.1 micro-mass actor for **a** rectangular, **b** trapezoidal, **c** triangular, **d** step profile rectangular using Ptolemy

for modeling, simulation, and design of concurrent systems [14]. In this chapter, to enhance the sensitivity of the sensor, four different shapes of the cantilever are used in the micro-mass optical MEMS sensor. The software codes for laser actor, photodetector actor, and four micro-mass actors using four different shapes of cantilever are developed using Ptolemy software [15]. Each mass actor consists of two optical fibers and one of the four shapes of cantilever like rectangular, trapezoidal, triangular, and step profile cantilever to make micro-mass actor. The developed micro-mass actors for different shapes of cantilever like rectangular, trapezoidal, triangular, and step profile are shown in Fig. 1.1.

1.3 Different Shapes of Cantilever

To enhance the sensitivity of the mass sensor, the study is carried out using two different polymers and four shapes of microcantilever. The polymers are chosen as microcantilever materials, because of its low Young's modulus, which gives more deflection than silicon. The polymer cantilevers can be easily fabricated. The fabrication cost also is lesser than silicon. But the polymers are temperature sensitive, so the cantilever should kept in a protective environment. The mathematical equation needed to develop the sensor and the micro-mass optical MEMS sensor for four different shapes are discussed below.

1.4 Rectangular-Shaped Micro-Mass Optical MEMS Sensor

The rectangular-shaped microcantilever as shown in block diagram Fig. 1.2a is connected between the two optical fibers. The micro-mass actor is developed for the rectangular-shaped cantilever using the open-source framework Ptolemy as shown in Fig. 1.2b.

Laser light that comes from the laser actor which acts as optical source of wavelength $\lambda = 850$ nm passes through two optical fibers separated axially. When the mass is added at the free end of the cantilever, the deflection will be in Y direction and by virtue of this deflection the output power detected at one of the fiber ends is varied continuously from maximum to minimum though the slit arrangement as shown below. This output power variation can be calibrated according to change in minute mass variation over the cantilever which in turn will constitute an accurate micro-mass optical MEMS sensor. The light coming out of the second optical fiber is detected by the photo detector. The model equation to make the micro-mass actor for rectangular-shaped cantilever was given below:

The modified stone's equation of the rectangular cantilever is given by

$$\delta = \frac{4(1 - \vartheta)\Delta\sigma}{E}\left(\frac{l_0}{t_0}\right)^2 \tag{1.1}$$

where E is Young's modulus, t_0 is thickness of the substrate, l_0 is length of the rectangular cantilever.

The fundamental resonant frequency (f_0) of the rectangular cantilever is given by

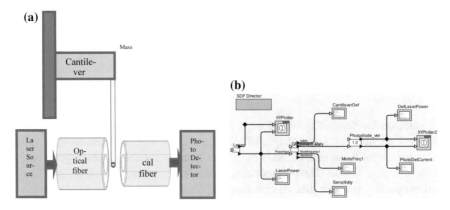

Fig. 1.2 **a** Micro-mass optical MEMS sensor using rectangular cantilever. **b** Micro-mass optical MEMS sensor using rectangular-shaped cantilever using Ptolemy (both Polyimide and Parylene)

$$f_0 = \frac{1}{2\pi}\sqrt{\frac{E}{\rho}\frac{t_0}{l_0^2}} \qquad (1.2)$$

where f_0 is fundamental frequency of the rectangular cantilever, E is Young's modulus, ρ is density of the cantilever material, t_0 is thickness of the substrate, and l_0 is length of the rectangular cantilever.

The sensitivity is the product of the deflection and fundamental frequency which helps us to find the minimum mass measured by the sensor.

The sensitivity factor of the rectangular cantilever as

$$\delta \cdot f_0 = \frac{2(1-\vartheta)\Delta\sigma}{\pi\sqrt{E\rho}}\frac{1}{t_0} \qquad (1.3)$$

where δ is deflection of the cantilever, ν is Poisson's ratio, $\Delta\sigma$ is stress or force applied on the cantilever, E is Young's modulus, ρ is density of the cantilever material, and t_0 is thickness of the substrate.

1.5 Trapezoidal/Triangular-Shaped Micro-Mass Optical MEMS Sensor

The block diagram of the trapezoidal and triangular micro-mass optical MEMS sensors is shown in Figs. 1.3a and 1.4a. The free end thickness of the trapezoidal cantilever is half the fixed end thickness as given in Eq. (1.6). The free end thickness of the triangular cantilever is one-tenth of the fixed end thickness. Trapezoid- and triangular-shaped micro-mass optical MEMS sensor model developed in Ptolemy is shown in Figs. 1.3b and 1.4b according to the model equation given below. The cantilever deflection, fundamental frequency, and sensitivity are measured for the two different shapes. Triangular-shaped micro-mass optical MEMS sensor is able to sense less maximum mass than trapezoidal-shaped optical MEMS sensor. The micro-mass actor for the trapezoidal/triangular is developed in the Ptolemy framework using the model equation given below:

The modified stone's equation of the triangular/trapezoidal cantilever is given by

$$\Delta\delta = \frac{8(1-\nu)\Delta\sigma l^2}{E(t_0 - t_1)^2}\left[\ln\left[\frac{t_0}{t_1}\right] + \frac{t_0}{t_1} - 1\right] \qquad (1.4)$$

The fundamental resonant frequency (f_0) of the triangular/trapezoidal cantilever is given by

$$f_0 = C\sqrt{\frac{S}{M}} \qquad (1.5)$$

The thickness of the free end of the cantilever trapezoidal cantilever is given by

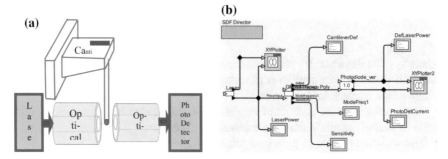

Fig. 1.3 Micro-mass optical MEMS sensor using trapezoidal cantilever. **a** Block diagram. **b** Model using Ptolemy (both polyimide and parylene)

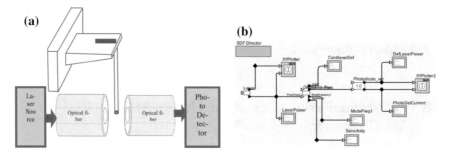

Fig. 1.4 Micro-mass optical MEMS sensor using triangular cantilever. **a** Block diagram. **b** Model using Ptolemy (both polyimide and parylene)

$$t_1 = \frac{t_0}{2} \tag{1.6}$$

The thickness of the free end of the cantilever triangular cantilever is given by

$$t_1 = \frac{t_0}{10} \tag{1.7}$$

The fundamental frequency of the triangular/trapezoidal cantilever is given by [16]

$$f_0 = C\sqrt{\frac{S}{M}} \tag{1.8}$$

where C is taper ratio whose value is 0.715, S and M are spring constant and mass of the cantilever.

1.6   Step Profile-Shaped Micro-Mass Optical MEMS Sensor

The spring constant of the cantilever is given by

$$S = \frac{c_3 E I_0}{l_0^3} \tag{1.9}$$

The mass of the cantilever is given by

$$M = \frac{l_0 A_0 \rho}{c_2} \tag{1.10}$$

The block diagram of step profile micro-mass optical MEMS sensor is shown in Fig. 1.5a. There are two sections in the cantilever. They are thick and thin sections, respectively. The thick section width is equal to the width of the fixed end of the rectangular cantilever. The thin section width is half of the fixed end width of the rectangular cantilever. The total length of the thick and thin sections is equal to length of the rectangular cantilever. Step profile-shaped micro-mass optical MEMS sensor model developed in Ptolemy framework is shown in Fig. 1.5b according to the model Eqs. (1.11)–(1.13). The cantilever deflection, fundamental frequency, and sensitivity are measured for the step profile-shaped micro-mass optical MEMS sensor which able to sense less maximum mass than rectangular-shaped optical MEMS sensor.

The moment of inertia of step profile rectangular cantilever is given by

$$I_{step} = \left\{ \frac{1}{(l_0 + l)^3} \left[\frac{4l_0^3}{I_0} + \frac{(3l_0 + 2l)l^2}{2I} \right] \right\}^{-1} \tag{1.11}$$

where I_0 and I, and l_0 and l are moments of inertia and length of the thick and thin sections of the step cantilever, respectively. Since thick and thin sections of the step design have equal width rectangular profile, their moment of inertias can be calculated from the basic expression $I = \frac{bt^3}{12}$.

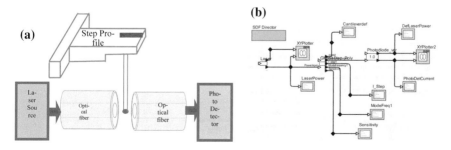

Fig. 1.5 Micro-mass optical MEMS sensor using step profile cantilever. **a** Block diagram. **b** Model using Ptolemy (both polyimide and parylene)

The Stoney equation for calculating surface stress-induced deflection of the step profile cantilever is given by

$$\Delta z = \frac{4(1 - \vartheta)\Delta\sigma}{E} \left[\frac{4l_0^3}{t_0^3} + \frac{(3l_0 + 2l)l^2}{2t^3} \right]^{\frac{2}{3}} \tag{1.12}$$

The fundamental frequency of the step profile cantilever is given by

$$f_0 = \frac{1}{2\pi} \sqrt{\frac{Et_0^3}{\rho(l_0t_0 + lt)(l_0 + l)^3}} \tag{1.13}$$

1.7 Results and Discussion

The range of the mass measured by the four micro-mass optical sensor models using four shapes are given in Table 1.1. The lowest mass is measured by triangular-shaped cantilever when compared to other shapes. Step profile rectangular can measure 70.1441% less mass when compared to rectangular cantilever. Triangular cantilever can sense 76.5217% less mass than trapezoidal cantilever is shown in Fig. 1.6.

The fundamental frequency of the rectangular cantilever is lowest, and the triangular is highest shown in Fig. 1.6. Sensitivity of the triangular shape is highest when compared to other shapes. Sensitivity of the step profile cantilever is 14.445% more than rectangular cantilever when Polyimide is used as cantilever material shown in Fig. 1.6. Parylene is used as cantilever material in the micro-mass optical MEMS sensor. The lowest mass is measured by triangular-shaped cantilever when compared to other shapes. Step profile rectangular can measure 70.1299% less mass when compared to rectangular cantilever. Triangular cantilever can sense 76.6067% less mass than trapezoidal cantilever which is shown in Fig. 1.7. The fundamental frequency of the rectangular cantilever is lowest, and the triangular is highest shown in Fig. 1.7. Sensitivity of the triangular shape is highest when compared to other shapes. Sensitivity of the step profile cantilever is 21.0181% more than rectangular cantilever when parylene is used as cantilever material shown in Fig. 1.7. Parylene

Table 1.1 Results of different shapes of cantilever

Micro-mass optical MEMS sensor		
Shape of the cantilever	Polyimide	Parylene
	Mass range	Mass range
Rectangular	101.9 µg–98.37 mg	101.9 µg–83.2 mg
Trapezoidal	101.9 µg–63.61 mg	101.9 µg–53.81 mg
Triangular	81.55 µg–28.39 mg	50.97 µg–23.996 mg
Step profile cantilever	203.9 µg–47.299 mg	50.97 µg–39.96 mg

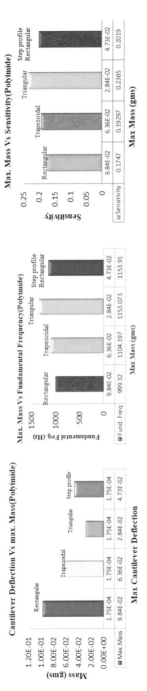

Fig. 1.6 Results of the optical MEMS sensor using different shape cantilevers (Polyimide)

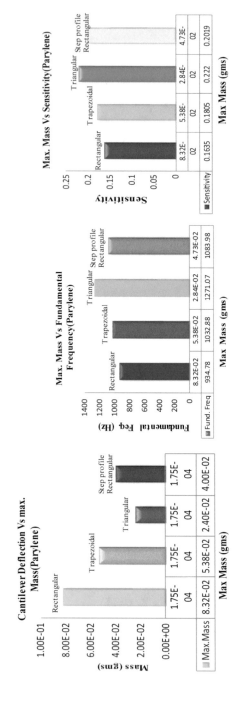

Fig. 1.7 Results of the optical MEMS sensor using different shapes cantilevers (parylene)

cantilever can sense approximately 15.4% less mass than Polyimide cantilever in all shapes. Parylene cantilever fundamental frequency is approximately 6.4% less than Polyimide. Parylene cantilever is approximately 6.2% less sensitivity than polyimide cantilever. The different shapes of the cantilever are modeled and simulated using the open-source framework Ptolemy. The deflection, fundamental frequency, and sensitivity of each shape were measured.

Parylene cantilever can sense approximately 15.4% less mass than polyimide cantilever in all shapes. Parylene cantilever fundamental frequency is approximately 6.4% less than polyimide. Parylene cantilever is approximately 6.2% less sensitivity than polyimide cantilever. The different shapes of the cantilever are modeled and simulated using the open-source framework Ptolemy. The deflection, fundamental frequency, and sensitivity of each shape were measured.

The triangular cantilever able to measure lowest mass among the four shapes, and fundamental frequency and sensitivity of the triangular is higher than other shapes. So far the system-level model has developed using Ptolemy framework.

1.8 Conclusion

The four micro-mass optical MEMS sensors are modeled using four different shapes of the cantilever in open-source framework Ptolemy. The deflection, fundamental frequency, and sensitivity of each shape are measured. The micro-mass optical MEMS sensor using triangular cantilever able to measure lowest mass among the four shapes and fundamental frequency and sensitivity of that sensor is high than other shapes. Two polymers are used as the sensor material for the cantilever. The detectable mass range measured by the triangular cantilever using polyimide as material is 81.55 µg–28.39 mg. The detectable mass range measured by the triangular cantilever using parylene as material is 50.97 µg–23.996 mg.

References

1. P.G. Waggoner, H.G. Craighead, Micro- and nanomechanical sensors for environmental, chemical, and biological detection. Lab Chip **7**(10), 1238–1255 (2007)
2. A. Gupta, D. Akin, R. Bashir, Single virus particle mass detection using microresonators with nanoscale thickness. Appl. Phys. Lett. **84**(11), 1976–1978 (2004)
3. M.Z. Ansari, C. Cho, J. Kim, B. Bang, Comparison between deflection and vibration characteristics of rectangular and trapezoidal profile microcantilevers. Sensors **9**, 2706–2718 (2009)
4. H.F. Hawari, Y. Wahab, M.T. Azmi, A.Y. Shakaff, U. Hashim, S. Johari, Design and analysis of various microcantilever shapes for MEMS based sensing. J. Phys. Conf. Ser. **495**, 1–9 (2014)
5. M. Abdel-Jaber, A. Al-Qaisia, M.S. Abdel-Jaber, Nonlinear natural frequencies of a tapered cantilever beam. Advanc. Steel Construct. **5**, 259–272 (2009)
6. S. Morshed, B.C. Prorok, Tailoring beam mechanics towards enhancing detection of hazardous biological species. Experiment. Mech. **47**, 405–415 (2007)

7. K. Yang, Z. Li, D. Chen, Design and fabrication of a novel T shaped piezoelectric ZnO cantilever sensor. Active Passive Electron. Comp. **2012**, Article ID 834961, 7 (2012)
8. M.Z. Ansari, C. Cho, Design and analysis of a high sensitive microcantilever biosensor for biomedical applications, in *International Conference on Biomedical Engineering and Informatics* (2008)
9. D.K. Parsediya, J. Singh, P.K. Kankar, Simulation and analysis of highly sensitive MEMS cantilever designs for "in vivo label free" biosensing. Proc. Technol. **14**, 85–92 (2014)
10. V. Gulshan Thakare, A. Nage, Design and analysis of high sensitive biosensor using MEMS. Int. J. Innovat. Sci. Eng. Technol. **2**(6), 697–701 (2015)
11. V. Mounika Reddy, G.V. Sunil Kumar, Design and analysis of microcantilevers with various shapes using COMSOL multiphysics software. Int. J. Emerg. Technol. Advanc. Eng. **3**(3), 294–299 (2013)
12. M. Chaudhary, A. Gupta, Microcantilever-based sensors. Defence Sci. J. **59**(6), 634–641 (2009)
13. C. Ptolemaeus, *System, Modeling and Simulation Using Ptolemy II, Creative Commons*. California, p. 10 (2014)
14. C. Brooks, E.A. Lee, X. Liu et al. *Heterogeneous Concurrent Modeling and Design in Java*. Technical Memorandum UCB/ERL M04/27, University of California, Berkeley
15. I. Mala Serene, M. Rajasekhara Babu, Z.C. Alex, Optical MEMS sensor for measurement of low stress using Ptolemy II. Advanc. Syst. Sci. Appl. **16**(3), 76–93 (2016)
16. J.A. Hoffman, T. Wertheimer, Cantilever beam vibrations. J. Sound Vibrat. **229**(5), 1269–1276 (2000)

Chapter 2
Enhancing the Performance of Decision Tree Using NSUM Technique for Diabetes Patients

Nithya Settu and M. Rajasekhara Babu

Abstract Diabetes is a common disease among children to adult in this era. To prevent the diseases is very important because it saves the human lives. Data mining technique helps to solve the problem of predicting diabetes. It has steps of processes to predict the illness. Feature selection is an important phase in data mining process. In feature selection when dimension of the data increases, the quantity of data required to deliver a dependable analysis raises exponentially. Numerous different feature selection and feature extraction techniques are present, and they are widely used filter-based feature selection method is proposed which takes advantage of the wrapper, Embedded, hybrid methods by evaluating with a lower cost and improves the performance of a classification algorithm like a decision tree, support vector machine, logistic regression and so on. To predict whether the patient has diabetes or not, we introduce a novel filter method ranking technique called Novel Symmetrical Uncertainty Measure (NSUM). NSUM technique experimentally shows that compared to the other algorithms in filter method, wrapper method, embedded method and hybrid method it proves more efficient in terms of Performance, Accuracy, Less computational complexity. The existing technique of symmetric uncertainty measure shows less computational power and high performance, but it lacks in accuracy. The aim of the NSUM method is to overcome the drawback of the filter method, i.e., less accuracy compared to other methods. NSUM technique results show high performance, improved accuracy, and less computational complexity. NSUM method runs in 0.03 s with 89.12% as accuracy by using Weka tool.

2.1 Introduction

Diabetes is the disorder that outcomes from lack of insulin in a human being blood. There are one more types of diabetes called is diabetes insipidus. When the patient mentions "diabetes," they mean diabetes mellitus (DM) [1]. A human with diabetes mellitus is called "diabetics." Diabetes symptoms include frequent urination, increased hunger, and thirst. If this is untreated, it will lead to serious complications. This complication includes kidney failure, stroke, damage to the eyes, heart disease, and foot ulcers. If there is a decrease in the sugar level in the blood, it will be called as a pre-diabetes [2]. Diabetes is causes when the pancreas does not secrete enough insulin to the body. Diabetes mellitus is of three types, namely type I diabetes mellitus, type II diabetes mellitus, and gestational diabetes. This is explained in detail. Type I diabetes is caused when the pancreas fails to yield enough insulin. It is referred as IDDM which is "insulin-dependent diabetes mellitus" alias "juvenile diabetes." The root cause is unknown. It will affect the people from below 20 years of age. It will continue throughout their life. They should follow strict diet and exercise. Type II diabetes starts when the insulin stops working in the human body. When the disease increases, the insulin level will be reduced. This is called as NIDDM non-insulin-dependent diabetes mellitus. The reason for this diabetes is obesity and lack of exercise. Type III diabetes is called as gestational diabetes. This will occur during the pregnancy. In the history of the patient, it will develop high sugar level. The recent research shows that 18% of women get this kind of diabetes during their pregnancy [3]. Based on the above understanding, diabetes should be controlled and predicted using predictive model technique. The research conducted in the USA in 2011 states that 8.3% of people have diabetes. It is the seventh leading reason for the death in the USA. It not only causes death but it also induces kidney failure, heart stroke, and blindness. People aged above 65 are affected much by diabetes. To prevent human death, it is important to predict the diabetes [4]. Centers of disease has given a statistics as shown below: [5]. 26.9% of the population affected by DM and age is above 65. 11.8% men affected by DM with age 20 or above and 10.8% women affected by DM with age 20 or above. Data mining methods promote healthcare researchers to extract knowledge from huge and complex health data. With the help of information technology, data mining delivers a valuable strength in diabetes research, which pointers to progress healthcare delivery and increase decision-making and enhance disease management [6]. Data mining techniques comprise pattern recognitions, classification, clustering, and association. Diabetes is important topics for medical research due to the durability of the diabetes and the massive cost from the healthcare suppliers. Primary noticing of diabetes eventually decreases the cost on healthcare suppliers for considering the diabetic patients [7–9], but it is a thought-provoking task. Detecting of diabetes, scientists can use DM people medical data and convert raw data into significant information by using data mining methods such as NSUM to construct an intelligent predictive model. It is commonly predictable that a large number of features can badly affect the performance of machine learning algorithms. Data mining has steps to extract the useful information from large datasets.The pre-processing of data takes important place to improve the model accuracy and performance. In machine learning, selecting a correct variable or attribute is known as feature selection. Usually, the feature selection technique is defined as the removing of the redundant and irrelevant feature, and these features will not be helpful in building a model.

These features should be helpful to accurately predict the outcome of the improved performance. It has been classified into four types, namely filter, wrapper, embedded, and hybrid.

A. *Filter Method*

Filter method is independent of the building model during the time of execution. It has a very low computational cost that is an advantage but the accuracy of the model will be produced is not be promised [10].

B. *Wrapper Method*

Wrapper method is dependent on building a model during classification or clustering based on the selected attributes [11]. It produces the high accuracy, but the computational complexity is too high [12].

C. *Embedded Method*

Embedded technique uses the FS as a portion of the training process. This method comparatively produces less accuracy than the wrapper method [13].

D. *Hybrid Method*

Hybrid technique is a mixture of wrapper and filter methods to achieve more accuracy, low computational time, and less cost [14] (Fig. 2.1).

Fig. 2.1 Feature subset selection process

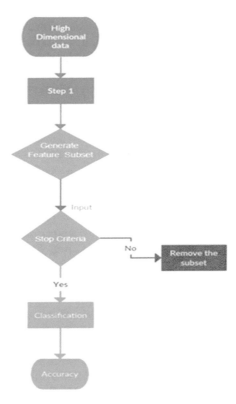

The aim of this paper is to improve the performance of the filter algorithm by using symmetrical uncertainty measure (SUM). We proposed a novel algorithm for SUM technique which is called as NSUM.

2.2 Related Work

Kohonen or SOM technique map the machine-learning tool which is used for heterogeneous data by providing unsupervised or supervised learning model [15–17]. It conveys the high-dimensional data to be more meaningful by identifying the similarities. This article compare the Random Forest and C4.5 algortihm which belongs to decision tree and SOM using hospital database. The datasets are obtained from Ministry of National Guard Health Affairs (MNGHA), Saudi Arabia. These datasets are collected from the three biggest regions in Saudi Arabia. The dataset collected by the author is from the below-listed hospital [18]. King Abdulaziz Medical City (SANG) in Riyadh, Central Region; King Abdulaziz Medical City in Jeddah, Western Region; Imam Abdulrahman Al Faisal Hospital in Dammam, Eastern Region; and King Abdulaziz Hospital in Alahsa, Eastern Region. The involvement of this learning is developing the data mining techniques to build an intelligent predictive model with real healthcare data which are extracted from hospital information systems by 18 risk factors. Makinen et al. worked on SOM technique to identify the association between the complications and the risk factors. Unsupervised machine learning technique is applied to healthcare profiles. 7×10 grid map units along with Gaussian neighborhoods method were applied to present the similarities and difficulties among variables [8]. Tirunagari et al. applied SOM cluster to reduce the dimensionality of the data by placing the patients in groups by using U-Matrix. The result of the analysis conveys that the patient who wants self-management was grouped Properly [9]. CFS technique assesses the feature ranking for subset attributes rather than whole attributes.CFS is done by the theory which is a good subset attribute contains high correlated with the target value, But not related to each other [19].

FCBF technique is a filter-based method of feature subset selection which identify the redundant attributes and irrelevant attributes without correlation analysis. Using cluster analysis, the subset selection is performed in combination of three ways [20]. This is called FS before clustering, FS after clustering, and FS during clustering's before clustering applies unsupervised FS methods during the preprocessing phase. It shows three different dimensions. They are irrelevant attributes, performance task in efficiency, and unambiguousness. This dimensions are used to improve the performance of the model [21]. FS selection during clustering applies genetic algorithm samples for heuristic-based search which uses fitness values to get the optimal result. On the output result, optimal k-means clustering is applied [22]. FS selection after clustering is applied a special metric technique Barthelemy–Montjardet distance first then it applies the feature selections. The hierarchical method generates cluster tree which is called as dendrogram [23].

2.3 Mutual Information

To measure the correlation between two or more attribute, mutual information (MI) is applied in the data mining process. MI is helpful to measure the features are correlated or not. It is designed as different among the sum of marginal entropy and joint entropy. MI value is zero for two independent features [24]. MI is helpful in feature selection so that good accuracy is obtained by building a classification model. This paper talks about Shannon's entropy. The High Dimensional data is D, N is the number if rows are the no of the attribute then D is defined as $D = M \times N$. Consider X, Y are the two random attributes then Probability density is defined as below equation [25].

$$\text{MI(X, Y)} = \sum_X \sum_Y p(x, y) \log \frac{p(x, y)}{p(x) * p(y)} \tag{2.1}$$

$$H(X) = -p(x)\log \int(p(x))dx \tag{2.2}$$

$$H(Y) = -p(y)\log \int(p(y))dy \tag{2.3}$$

$$\text{MI(X, Y)} = H(X) - H(X, Y) \tag{2.4}$$

$$MI(X, Y) = H(X, Y) - H(X|Y) - H(Y|X) \tag{2.5}$$

2.3.1 Symmetric Uncertainty

Symmetric uncertainty is used to measure the fitness measure for the selected feature and target class. The feature that has the high value will get high ranking and importance. Symmetric uncertainty is defined as follows:

$$\text{SU(x, y)} = (2X \text{ MI (X, Y)})/(H(X) + H(Y)) \tag{2.6}$$

H(X)—The entropy of a random variable. The probability of X random variable is P(X).

It is calculated by Eq. 2.2.

H(Y)—The entropy of a random variable. The probability of X random variable is P(Y).

It is calculated by Eq. 2.3.

MI partiality features has huge number of different values and regularizes within range of [0, 1].

The SU(X, Y) shows that knowledge of the object value strongly represents. The values of other than the SU(X, Y) value 0 indicates the independence of X and Y.

2.3.2   Proposed Algorithm

Symmetric uncertainty ranking-based feature selection
Input dataset—(f1, f2, f3 … fn, C), threshold λ.
Output dataset—an optimal subset of features.

1. Begin the algorithm.
2. Calculate the SUi for each feature fi.
3. Check SUi is greater than the λ value.
4. Store fi into D' array variable.
5. Find the length of the D' array.
6. Calculate the middle index (MID). For D'.
7. Select the first value in the pointer. Travel (first point to the midpoint).
8. Select the last value in the pointer. Travel (last value point to the midpoint).
9. Using the temporary variable, swap the data according to the ranking. First part.
10. Using the temporary variable, swap the data according to the ranking. Second part.
11. /*** {Select the First feature! = Midpoint} Sort the data according to the feature.
12. Variable array 1 = Store the sorted data.
13. /** {Select the last feature! = Midpoint} Sort the data according to the feature.
14. Variable array 2 = Store the sorted data.
15. Add the variable array 1 and variable array 2.
16. OUTPUT is optimized features.

2.4   Experimental Result and Discussion

In our new work, we evaluated the efficiency of the recommended technique. The aim of our plan is to assess the method in terms of speed, no of selected attributes, predictive accuracy for a J48 classifier selected feature. The algorithm matched in contradiction of some previously existing techniques: SOM, chi-square, relief, and FCBF on the diabetes high-dimensional datasets. NSUM approach outcome is less number of features as compared to SOM, FCBC, and Relief, grades in the reduction of time for the resultant mining algorithm. A list of datasets used in our approach is from the UCI repository [26]. A brief summary of datasets is described in Table 2.1.

Table 2.1 Feature technique run time

Technique	Time (ms)	Correctly identified instances	Incorrectly identified instances
SUM	0.06	79.08	20.92
NSUM	0.03	87.12	12.88

Fig. 2.2 Performance of
NSUM algorithm

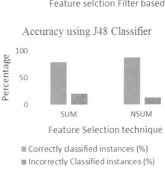

Fig. 2.3 Accuracy of the
techniques

2.5   Conclusion and Future Scope

Decision tree data mining technique is used to help healthcare specialists in the diagnosis of diabetes millitus disease. Applying health mining is helpful to healthcare, disease diagnosis, and treatment. The future scope will be using a hybrid model increase the accuracy and performance optimization (Figs. 2.2 and 2.3).

References

1. S. Siddiqui, Depression in type 2 diabetes mellitus—a brief review. Diabetes Metab. Synd. Clin. Res. Rev. **8**(1), 62–65 (2014)
2. K. Rajesh, V. Sangeetha, Application of data mining methods and techniques for diabetes diagnosis. Int. J. Eng. Innov. Technol. (IJEIT) **2**(3) (2012)
3. S. Sarma Kattamuri, Predictive modeling with SAS enterprise miner: practical solutions for business applications (SAS Institute, 2013)
4. W. Gregg Edward et al., Association of an intensive lifestyle intervention with remission of type 2 diabetes. JAMA **308**(23), 2489–2496 (2012)
5. A.R. Mire-Sluis, R.G. Das, A. Lernmark, American diabetes association. Diabetes/Metab. Res. Rev. **15**(1), 78–79 (1999), http://www.diabetes.org
6. I. Yoo et al., Data mining in healthcare and biomedicine: a survey of the literature. J. Med. Syst. **36**(4), 2431–2448 (2012)
7. R. Li et al., Cost-effectiveness of interventions to prevent and control diabetes mellitus: a systematic review. Diabetes Care **33**(8), 1872–1894 (2010)
8. J.-H. Lin, P.J. Haug, Data preparation framework for preprocessing clinical data in data mining, in *AMIA Annual Symposium Proceedings* (American Medical Informatics Association, 2006)
9. M. Luboschik et al., Supporting an early detection of diabetic neuropathy by visual analytics, in *Proceedings of the EuroVis Workshop on Visual Analytics (EuroVA)* (2014)

10. V.-P. Mäkinen et al., Metabolic phenotypes, vascular complications, and premature deaths in a population of 4,197 patients with type 1 diabetes. Diabetes **57**(9), 2480–2487 (2008)
11. S. Tirunagari et al., Patient level analytics using self-organising maps: a case study on type-1 diabetes self-care survey responses, in *2014 IEEE Symposium on Computational Intelligence and Data Mining (CIDM)* (IEEE, 2014)
12. P. Xing Eric, M.I. Jordan, R.M. Karp, Feature selection for high-dimensional genomic microarray data. ICML **1** (2001)
13. N. Hoque, D.K. Bhattacharyya, J.K. Kalita, MIFS-ND: a mutual information-based feature selection method. Expert Syst. Appl. **41**(14), 6371–6385 (2014)
14. S. Das, Filters, wrappers and a boosting-based hybrid for feature selection. ICML **1** (2001)
15. D. Ballabio, M. Vasighi, P. Filzmoser, Effects of supervised self organising maps parameters on classification performance. Anal. Chim. Acta **765**, 45–53 (2013)
16. R. Wehrens, M. Lutgarde, C. Buydens, Self-and super-organizing maps in R: the Kohonen package. J. Stat. Softw. **21**(5), 1–19 (2007)
17. D. Wijayasekara, M. Manic, Visual, linguistic data mining using self-organizing maps, in *The 2012 International Joint Conference on Neural Networks (IJCNN)* (IEEE, 2012)
18. T. Daghistani, R. Alshammari, Diagnosis of diabetes by applying data mining classification techniques. Int. J. Adv. Comput. Sci. Appl. (IJACSA) **7**(7), 329–332 (2016)
19. M.A. Hall, Feature selection for discrete and numeric class machine learning (1999)
20. L. Yu, H. Liu, Feature selection for high-dimensional data: a fast correlation-based filter solution, in *Proceedings of the 20th International Conference on Machine Learning (ICML-03)* (2003)
21. L. Talavera, Feature selection as a preprocessing step for hierarchical clustering. ICML **99** (1999)
22. L. Boudjeloud, F. Poulet, Attribute selection for high dimensional data clustering. ESIEA Recherche, Parc Universitaire de Laval-Change 38 (2005)
23. R. Butterworth, G. Piatetsky-Shapiro, D.A. Simovici, On feature selection through clustering, in *Fifth IEEE International Conference on Data Mining* (IEEE, 2005)
24. G. Qu, S. Hariri, M. Yousif, A new dependency and correlation analysis for features. IEEE Trans. Knowl. Data Eng. **17**(9), 1199–1207 (2005)
25. H. Almuallim, T.G. Dietterich, Learning with many irrelevant features. AAAI **91** (1991)
26. K. Bache, M. Lichman, UCI machine learning repository, http://archive.ics.uci.edu/ml (University of California, School of Information and Computer Science. Irvine, CA, 2013)

Chapter 3
A Novel Framework for Healthcare Monitoring System Through Cyber-Physical System

K. Monisha and M. Rajasekhara Babu

Abstract In recent years, the major concern with people is healthcare. Humans are susceptible to various chronic diseases such as diabetes insipidus, kidney diseases, and eating disorders. The patient suffering from the mentioned diseases should be monitored and treated regularly to avoid any serious conditions. Thus, an embedded technology is developed to transfer the patient's health information through sensor to network and then to the cloud storage. The existing technologies usually monitor the patient's clinical data and share the sensor data to cloud. But, the system does not perform any data analysis or actuation process for efficient remedial treatment. In critical situations, the patient also requires the doctors and clinical assistants to be alongside to provide treatment immediately. Therefore, it requires a smart improvement in the current technology. In our methodology, we implement cyber-physical system (CPS) technique for healthcare system. CPS technology classifies the implementation into three parts, namely communication, computation, and actuation or control. CPS continuously monitors the patient's health parameters such as blood glucose (BG) level, blood pressure (BP) level, body temperature (BT) level, and heart beat (HB) rate. When the health parameter value reaches their critical bound, then through actuators the patients are treated inevitably as a remedial measure. The proposed system benefits the patients, doctors, and clinical assistants in reducing the overhead of assisting all the patients during the inconvenience period. Due to increased physical connectivity constraints, embedded systems and networks have more security exposures. Especially in healthcare systems, the lack of importance on device security has headed to numerous cyber-security gaps. Therefore, a proper investigation is needed on the CPS security issues to make sure that systems are working safe. Furthermore, security resilience and robustness are discussed. Finally, some healthcare data security challenges are elevated for the future study. The proposed CPS model decreases the overhead of medical representatives. This approach also decreases the time and cost complexity compared to the previous works.

© The Author(s), under exclusive license to Springer Nature Singapore Pte Ltd. 2019
P. V. Krishna et al., *Internet of Things and Personalized Healthcare Systems*,
SpringerBriefs in Forensic and Medical Bioinformatics,
https://doi.org/10.1007/978-981-13-0866-6_3

3.1 Introduction

Cyber-physical system (CPS) is a recent research topic that has received widespread attention across different domains, including smart grids, smart hospitals, smart house, and energy [1]. Generally, CPS is a technology comprised of 3Cs namely communication, computation, and control, respectively. CPS forms closed loop by connecting machine to machine where the physical components (e.g., sensors and Wi-Fi boards) actively interact with the cyber-space network (e.g., Internet) for transmission of data to cloud storage and response back to the physical space using actuator machine. CPS mutates the process of interaction occurring in the corporal world, as distinct device needs different form of safety levels built on robustness of control system and the sensitivity of the data that is exchanged. CPS has challenges in preserving the security and privacy in each application. In healthcare system [2], it is important to facilitate the security, privacy, reliability, and assurance for effective health device communication. Thus, implementing an efficient healthcare system using CPS requires a secure pervasive network model. A model, namely Pervasive Social Network (PSN) for healthcare-based system, is promoted for sharing the patient's health data collected from the medical sensor over the secure network [3, 4]. The data used in e-healthcare application is represented as Electronic Healthcare Record (EHR) which has a vital scope in improving the healthcare usability, human experience, and data intelligence. It is observed that EHR could eventually store huge volume of data allowing effective retrieval of clinical records [5].

In healthcare system, sharing of patient's medical record should help in making the user experience smarter, better endorsement for both doctors and patients, understanding the data patterns and diseases to provide better healthcare quality service. Due to the significant feature of CPS, its application is used everywhere with testified results. The Health CPS also has a prompting growth due to the evolution of hardware techniques having the standard bandwidth by integrating the intellectual radio-based networks to disclose the utilization range of frequency band. Machine-to-machine communication in CPS, wireless sensor network (WSN), and cloud computing has become a fundamental part for any Internet-based applications [6]. As many parts included in the IoT applications, it also required to have a concern on security problems relating to WSN, CPS, and cloud computing. The security concerned problems can appear in the background of IoT or CPS having Internet protocol standard for connectivity. Consequently, in recent surveys many effects have been taken to handle the security issues in the CPS model. Different security approaches are followed namely, providing security to only particular layer and in some cases providing end-to-end security to the CPS applications [7, 8]. Relating to the security issues in the CPS network layer, it is observed that more than many thousand health consumer devices were in consent to distribute spam mails, brute force attack, and other outbreaks. Thus, it is required to provide the security for the entire application starting from physical devices adding with the network and ending with the cloud storage. There should be a separate team working on these security concerns. For example, in an application, namely smart hospital setting, the information technology (IT) crew has the entire control of the network module including the IoMT devices with an IP address and endpoint devices. But in such scenario, it is an impractical act to expect the IT team may be familiar with the context of each device connected to the wireless network even in the case if the team has the privileges to load patches or access the devices remotely.

Hence, many security aspects have been applied to overcome the security threats in the healthcare application; one such security aspect or technique is known as blockchain technology. Blockchain technology facilitates in secure sharing of CPS datasets among practitioner groups, researchers, and other shareholders. Blockchain technology is initially used for observing and recording all the financial transactions which occurs virtually in online, for example, cryptocurrency and bitcoin system [9, 10]. Thus, employing blockchain affords transparent transactions and easy tracking/detection of any modifications in the system. Blockchain consequently ensures that it can be applied to enhance the security in CPS or any online transaction process. Blockchain also ensures in preserving integrity among the participants in the transaction hub, while the datasets are shared across the network. To maintain the integrity among the datasets, blockchain applies a Reference Integrity Metric (RIM) for the CPS datasets. RIM checks for the integrity whenever the datasets are shared or downloaded in the system's hub [11, 12]. For more information, blockchain maintains a centralized hub which stores the participant repositories as references where the datasets are warehoused and distributed. The participant's record such as owner information, sharing strategy, and address are stored as blocks and shared by all the members in the hub [13]. Different data structures are used for sharing the healthcare records which prevent the compatibility and bound data comprehension due to disparate use of physiological parameters. Semantics and structure can also be settled upon, but data consistency, privacy, and security are a big concern. Cyber-attackers target on authority benefactors and centralized database storage [14, 15]. It enables a consistent view with the patient healthcare record across data proliferating network which leads to a problem. Hence, blockchain methodology is applied for sharing patient's health information. Blockchain approach promotes decentralized or single centralized database for recording all transactions in favor of trust of network consent with corroboration of semantics and system interoperability.

3.2 Related Work

3.2.1 Wireless Body Area Network (WBAN) in Healthcare System

In [16] methodology, a wireless body area network (WBAN) is designed for implementing healthcare application. WBAN uses the clinical band to transfer the patient's physiological parameters from the sensor node through microcontroller using wireless communication system. To increase the synchronization aspect between the sensor node devices and other network node devices, clinical bands are introduced to reduce the interventions at different health centers [17, 18].

The proposed system employs the multi-hoping method to transfer the collected sensor data from one location to another isolated location using wireless gateway board. The exchange of information happens by connecting the sensor node to the Wi-Fi node or local area network (LAN). The proposed WBAN for medical applications ensures in facilitating the health centers, doctors, and clinical assistants to access the patient's physiological parameters at anywhere through both offline and online [19, 20]. The defined methodology also reduces the medical cost, human faults, and periodical checkup for patients attended by medical professionals. In [21], WBAN security and privacy aspects are discussed. In smart technologies, it is important to provide a high-level security and privacy which is a vital scope for healthcare monitoring applications. Healthcare monitoring system is responsible for observing and transferring the patient health data over the network to the cloud for storage purposes. Hence, it is essential to protect the health data parameters from the intruder's exploitation. Therefore, the proposed system works in deploying the WBAN based on the privacy and security aspects [22, 23]. The WBAN communication architecture is also discussed with security and privacy threats that occur while integrating the hardware components (e.g., sensors and microcontrollers) with software development environment (e.g., cloud and network topologies). In the proposed work, it is concluded that the security threats, audit trails, and privacy challenges of healthcare application are described within the legal framework for further awareness. In [24], the experimental setup focuses on data gathering protocol or convergecast in WBAN for healthcare applications. The contribution starts with evaluating the effect of postural body movement along with the various multi-hop data gathering protocol approaches [25]. The system also evaluates the performance of delay-tolerant network (DTN) and wireless sensor network (WSN) through substantial stimulators. The simulations are executed using the OMNet++ simulator improvised with MiXiM structure and WBAN realistic network protocol. Two strategies are used, namely gossip-based strategy and multi-path-based strategy, for WBAN improvisation [26, 27]. Multi-path-based strategy represents virtuous dynamic performances, while gossip-based strategy presents a proper reliability for the healthcare system. An innovative hybrid convergecast strategy is experimented for better consent in terms of agility, day-to-day delay, and energy utilization.

3.2.2 Electronic Health Record (EHR) Assisted by Cloud

In [28], Clinical Document Architecture (CDA) is generated and integrated with health records for secure exchange of information using cloud computing. It is noted that electronic health record is used for storing the patient's physiological parameters. Hence, prerequisite of interoperability is required for deploying EHR for an improvised patient healthcare and security. CDA is a fundamental document standard developed by HL7 for interoperability concept between heterogeneous domains.

The broadcast of CDA document standard is crucial for interoperability format. It is also noticed that many hospitals showed less interest to acquire CDA document format due to its cost consent and maintenance used for deploying the software for interoperability. The stated drawbacks are reduced in the proposed method [29, 30]. In the proposed system, CDA document generation and integration built on cloud computing is realized with an OpenAPI service. The proposed system ensures that hospitals can generate CDA document properly without in the need of procuring and installing the software. The defined CDA document integration model also generates multiple documents for a single patient and integrates all the patient documents into single CDA document. The single CDA document for each patient enables doctors, clinical assistants, and hospitals to acquire the medical information in sequential order [31]. The CDA integration model assists in providing interoperability between hospitals and quality of patient care. Along with the benefits, CDA integration document also reduces cost and time to be spent on data format adaptability. In [32], it is observed that there is advancement in information technology which benefits many domains in their technical progress. One among the benefited domain is e-healthcare system improvised with new technologies. Consequently, the technology adopted by the healthcare system results in handling huge volume of clinical data. The data obtained from the various IoT devices is generated in a very short span of time, which eventually makes difficult in accessing the data. The problem also gets more complicated with the database structure after storing the records. In order to provide a novel healthcare service model, cyber-physical system is proposed to enhance the quality-oriented service for health centric applications [33]. In addition, the e-Health CPS is facilitated by implementing on cloud and big data technologies for better analytical purpose.

The proposed Health CPS model consists of three layers, namely (a) data collection layer—It consists of all e-health standards integrated together as a collection; (b) data service-oriented layer—This layer is responsible for providing all Health CPS-related services; (c) data management layer—The management layer controls parallel computing and distributed data storage in healthcare system. Finally, it is observed that a smart healthcare system is implemented using cloud and big data technologies. In [34] proposed system, the electronic patient centric records are handled and stored in cloud using a secure role-based technique. In recent observations, cloud technology encounters a rapid growth applied in different applications. Eventually, a cloud server has to adapt a larger data storage with increasing popularity in smart technologies such as smart hospital, smart grid, smart city, and smart energy. Hence, many hospitals started to store patients' record in an electronic form rather than manual data. These electronic health records are stored through cloud-based mechanism for better retrieval of data and quality of service [35]. However, despite cloud having the advantage in storage, it also has the issues in security aspect involving to unauthorized users. A cryptographic technique, namely role-based encryption model, is implemented to frame a secure and flexible cloud-based system to store electronic health records. The role-based encryption system ensures in framing the policies in the cloud system by avoiding the unauthorized user access [36, 37]. The proposed role-based encryption (RBE) system also establishes the security and reliability consent with Personally Controlled Electronic Health Record (PCEHR) system developed by the research center. Thus, the implemented system has the capability to deploy its role-based accessible secure method in any healthcare related applications. The methodology also observes experimental access procedures based on the roles and delivers secure storage access in cloud server imposing these access specific strategies.

3.2.3   Data Security in Healthcare Application

In [38] survey, big data technology has become a driving factor for many applications such as healthcare research, information technology, and educational institutions [39, 40]. Big data technology has many advantages such as time and cost reduction, and advanced product development. However, big data technology also encounters many challenges and impediments in providing security, privacy, and proficient talents in software development. One among those applications in big data is e-healthcare system where the health records are most susceptible to the attackers. Those attackers can easily find out the sensible data and spread them across the network which eventually leads to data breach [41, 42]. Hence, authentication is an important aspect in the healthcare system to protect those sensitive data from breaching by using various techniques such as.

3.2.3.1   Data Encryption

Encryption allows protecting the ownership of the data by avoiding any unauthorized user access to database. Encryption algorithms such as RSA, DES, RC4, AES are used as an encryption scheme for any efficient data privacy management.

3.2.3.2   Authentication

It involves authenticating the users to access the e-healthcare records by applying cryptographic protocols such as secure socket layer (SSL) and transport layer protocol (TLP).

3.2.3.3   Access Control

When an authenticated user accesses the e-health system database, they are regulated by the access control policy though the user is authenticated. Here the user gets their rights and privilege only when they are authorized as patients. Some of the techniques used for access control are sequence access control (SAC) and role-based control (RBC).

3.2.3.4 Data Masking

Masking involves hiding of sensitive data with an unidentifiable string. But, this method does not identify the original data after masking as such in encryption algorithm. But it uses some unique strategy for decrypting the data (which is encrypted) into original data values such as patient name, blood group, date and time the patient diagnosed with sickness.

3.3 Framework for Healthcare Application Through CPS

Healthcare system requires a constant improvisation in its organization resources and structure. Accordingly, many health research organizations manage in improving the efficiency and reliability of Electronic Health Records (EHR). The medical institutions improved their proficiency through unification adapters and health monitoring devices over the network module. These organizations also make an operable function over the influenceable variables cached in their healthcare server [43]. However, the operations defined in the server defect in their vital extensions as the structure of the healthcare system is more complex than the predicted one. The modifications that happen frequently or rarely in the server frameworks can affect the service delivered by the wellness program. The changes can affect the service standards by performing in an unusual behavior. For example, a doctor or medical assistants will be unable to provide proper treatment to patients in given time due to irregular update along with unexpected costs. Hence, a smart system is required by integrating the service-oriented cloud with other smart solutions to monitor the patients regularly. The patient's heath parameters are observed by sensors, microcontrollers, and other smart devices such as computers and mobiles. The interconnected solutions are accessible to clinical data which is presented through some algorithms and frameworks. The patterns are recognized through the algorithms for each patient with the responses stored in the data servers.

Thus, to provide the best solution to the healthcare care organizations, smart systems are employed. The smart system with effective machine-to-machine communication is provided through cyber-physical system (CPS). CPS framework is deployed for effectual healthcare monitoring system. CPS is a mechanism developed using problem-solving algorithms connected to the Internet users through network adapters. CPS is a technique built upon logically by merging the optimized algorithms with the networks and smart physical devices. CPS is employed in the platform whenever a smart implementation is required in an environmental application. In Fig. 3.1, a framework is designed for healthcare monitoring system by applying CPS notions. In the design, the framework is divided into three layers, namely (a) application layer—It consists of the applications defined for CPS technologies; (b) data layer—It includes entities or the members who analyze data for further concern in the system; (c) CPS layer—This layer consists of actual CPS implementation for smart hospital.

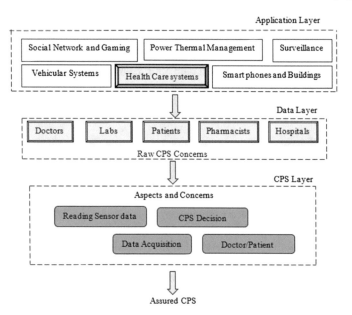

Fig. 3.1 Health CPS framework

Each layer in the defined framework makes vital scope for effective healthcare monitoring system through assured CPS. The objective is to provide the well-defined framework in coordination with common architectural standards in the scope of deploying the smart hospital. Application layer—It consists of the domains for the smart system, namely smart grid, smart hospital, smart energy, smart city, smart vehicle, and smart house. In the proposed system, smart hospital is implemented using CPS framework. Data layer—In this layer, members or entities to analyze the medical data are represented. The entities are patients, laboratories, doctors, pharmacists, and hospitals. The doctors and clinical assistants' analyze the data stored in cloud for providing the treatment to patients. This layer receives the assured and measured patient's health record.

CPS layer—This layer includes aspects and concern of smart hospital. The actual implementation is resided in this layer. The sensors are placed over the patient's body making each sensor area as a node. The sensor sends the physiological values to the microcontroller, thereby sending to the cloud storage. In cloud, decisions are made whether to provide treatment to patient or not based on the physiological parameters which are termed as CPS decision. Data acquisition happens when the doctor or any clinical assistants access the patient's data from cloud. After accessing, the doctors or nurses decide the kind of treatment to give to the observed patient. Thus, CPS enables an active interaction between the doctors and patients by enabling a proficient communication and computation model over the network. Hence, CPS provides an assured mechanism or algorithmic concept for implementing smart hospital.

3.4 Internet of Medical Things (IoMT)

Internet of Medical Things (IoMT) is a technology of connecting the IoT devices with Medicare application in the IT system through embedded networks [44]. IoMT applies the concept of machine-to-machine communication using the Wi-Fi-enabled devices. In IoMT, embedded devices transfer the health records over the computer networks and store the data in cloud for future analysis. IoMT includes remote monitoring of patient suffering from long-term or chronic diseases such as heart ailment, stroke, and diabetes. IoMT also tracks the patient's health conditions or orders, patient movement in the hospital or home, and patient's wearable e-health devices. IoMT collects the medical records and sends to the cloud for the caretakers to analyze the data.

The microcontroller or Wi-Fi-enabled device is connected to the data analytical dashboard and to the sensors equipped with patient's bed. These sensors and dashboards which observe the physiological parameters can be deployed as IoMT technology. IoMT comprises both software and hardware architecture which is used as a foundation for future low-power and wireless communication of wearable devices. These wearable devices are placed on the patient's body and communicate non-invasively through body tissues. IoMT allows the following features in healthcare system, namely (a) monitoring the patient remotely and storing the physiological parameters in cloud observed by wearable sensors; (b) controlling the actuators remotely deployed in the patient's body; (c) machine-to-machine communication enabling the system to function as closed-loop application. In Fig. 3.2, the basic concept of IoMT architecture is represented along with the components included to design the IoMT model. IoMT is described when the medical devices are compromised or connected to the IoT technology by framing as Internet of Medical Things. As standard IoT, IoMT also contains physical space consists of hardware boards and sensors, where the observed sensor data is transferred to the central database in the form of electronic health record. The EHR format of patient's physiological parameter allows for efficient monitoring of any remote sensing model. In the above IoMT representational diagram, it is explained that the data from the patient's wearable devices is exchanged to the database through Wi-Fi-enabled microcontrollers. The records are then stored in the central storage system such as cloud server. The data in the cloud server is stored in the form of Electronic Health Records having the standard health parameter values. The patient's health value stored in the cloud server is termed as Health Cloud allowing for remote monitoring and sensing of patient's health condition at anywhere. Remote monitoring also includes remote control actuators, remote measuring of physiological parameters, and cloud storage having electronic medical records.

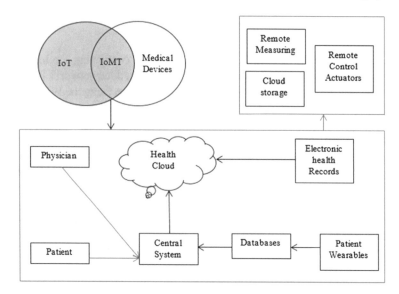

Fig. 3.2 IoMT basic architecture

3.5   Proposed Method

In general, resource sharing means sharing the metadata (hence the receiver can recognize the resource). Once a part of resource is accessed, it could be saved for future study. Here the challenging issue is data privacy, for example, permitting the user to analyze their resources. If our primary goal is about data processing, we would have adopted a secured data processing model. Sharing of medical data is vital for research and progress in healthcare services. But medical data values are dispersed in various healthcare systems. It is considered as the valuable information for any researcher to proceed for experimentation and analysis. It is obvious for any individual to own and control their medical data for privacy issues. Our proposed architecture allows this by using blockchain platform as storage system. Based on our architecture, the resource data is well organized and saved properly to avoid data loss. An interleaved memory-based cloud data storage system in blockchain platform leads to efficient cloud data storage and access. In our proposed approach, Organized Cloud Data Storage (OCDA), the data is preserved by dividing the data into chunks and sending it to different cloud storage systems. The fundamental aspect of data storage should be single and reliable.

While in cases like medical records and financial transactions, it is better to have a single version. Instead of holding multiple iterations, it is better to have a single copy for good economic reasons. The blockchain approach holds single copy of data which is distributed between the users. Hence, we adapted the interleaved memory approach for distributed and organized data storage as displayed in Fig. 3.3. The data is divided and distributed into four different storage systems.

Fig. 3.3 DATA-READ
representation of OCDA

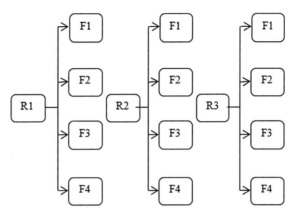

Fig. 3.4 Example of
OCDA—initial phase

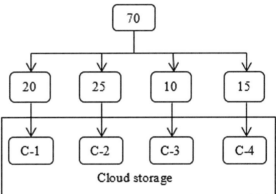

$$\sum_{i=1}^{n} V_i = \{V_1, V_2, V_3, \ldots, V_n\} \tag{3.1}$$

$$\sum_{i=1}^{n} C_i = \{C_1, C_2, C_3, \ldots, C_n\} \tag{3.2}$$

where V is the data value and C is cloud storage system.

$$((R1, V1, C1), (R1, V2, C2), (R1, V3, C3), (R1, V4, C4))$$

where R is Read, V is value, and C is cloud.

For example, let us consider data value is 70; it is divided as $70/4 = V1$, $(70 - V1)/3 = V2$ and stored into four different storage systems as displayed in Fig. 3.4.

Consider there are two datasets V_i and V_k. Initially, the number of data elements present is organized as

$$n(V_i \cup V_k) = n(V_i) + n(V_k) - n(V_i \cap V_k) \tag{3.3}$$

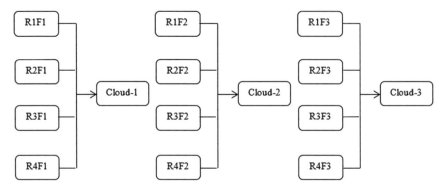

Fig. 3.5 OCDA approach

Next, the required number of cloud storage is calculated according to the datasets as follows.

$$C^i \cdot C^k = \Delta C^{i+k} \tag{3.4}$$

where C^i and C^k are number of cloud storage systems required for data i and k. Though the approach is working well, our motive is to perform it in an organized way. Hence, the following technique as displayed in Fig. 3.5 is designed for optimized and organized distributed cloud storage system. Here all data is organized in such a way that C1 stores all F1, C2 stores all F2, respectively.

3.6   Result and Discussion

To evaluate the use of OCDA approach, we need to evaluate the communication cost using the data collected through various sensors such as pulse rate sensor, temperature and humidity sensor, ultrasonic sensor, and blood pressure sensor. Intel Edison Arduino boards are used to measure the data transmission cost of virtual resources. The boards are connected to the same Wi-Fi network, and each board executes one resource at a time. At a time, each board communicates with the other by sending 1000 sequential requests. Once the request is sent, the resource must wait for the acknowledgment. Likewise, the resources wait and proceed with the future requests. Figure 3.6a, b displays the time taken to send the data of various sizes using OCDA approach. Before experimenting with the real-time data, let us test our approach with the available benchmark dataset. For our experimentation, we have obtained diabetes dataset from [45]. There are almost 20 data fields comprised of insulin dose, blood glucose measurement, hypoglycemic symptoms, and so on. Each data value is organized in such a way that one resource is communicated at a time. To avoid overloading, the resources are scheduled in a proper way and communicated. The data

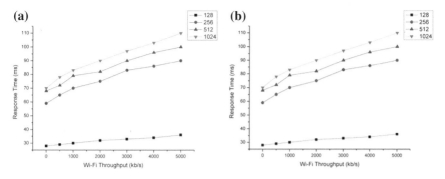

Fig. 3.6 a 8–64 bytes. **b** 128–1024 bytes

values present in the dataset are NPH insulin dosage, ultralente insulin dose, unspecified blood glucose measurement, undetermined blood glucose level, pre-breakfast blood glucose level, post-breakfast blood glucose level, pre-lunch blood glucose level, post-lunch blood glucose level, pre-supper blood glucose level, post-supper blood glucose level, pre-snack blood glucose level, hypoglycemic indications, typical meal digestion, more-than-usual meal digestion, less-than-usual meal digestion, typical workout activities, and unusual workout activities. These resources are distributed to different cloud based on the size. The below OCDA approach experiences slight delay as the data size increases, but it helps in avoiding overload of sensor data. Hence, our approach never experiences overload. Hence, OCDA is an optimal approach for storing virtual sensor resources. The OCDA classifies the data value and communicated to organized cloud storage system. This approach not only leads to secured transaction but also well-organized communication. Continuous resource communication sometimes leads to system overloading. But our approach classifies the number of resources and calculates the number of cloud storage systems required for it.

3.7 Conclusion

The main objective of implementing CPS is to monitor the patient suffering from chronic diseases effectively to overcome the severity in patient's health condition. The modeling also involves active observation of patient's physiological parameters such as body oxygen (BO) level, heartbeat (HB) rate, blood glucose (BG) level, and blood pressure (BP) level. The observed values are then uploaded to the cloud server and analyzed using some defined framework for determining the patient's body condition. CPS categorizes the implementation into three spaces, namely physical space—It consists of hardware components such sensors and microcontroller; cyber-space—It includes the actual computation where the sensor data that is transferred over the network is store in cloud for computing purposes; social interaction space—In this space, the actual interaction between machine and machine occurs, and it also involves the interaction between the patient and doctors or clinical assistants. In critical situations, the data analyzed is the cloud and fixes a status that if value is higher than the threshold value, then a notification is sent to the doctors or clinical assistants' mobile devices to ensure the patient's condition. Hence, it is observed that the patient's health condition data is very sensitive and important to handle while transferred over the network module. In this paper, a CPS framework is developed for remote monitoring of patient along with some security or safety measures that are also implemented to protect the electronic health records from cyber-attacks. In the proposed method, a novel OCDA approach is used for preserving the data by dividing the data into chunks or blocks and sending it to different cloud servers. The proposed architecture uses the concept of blockchain methodology for distributed storage system. The approach uses the concept of distributed data storage with the perception of single data server as similar to interleaved memory. The objective is to split the sensor data key value and store the key values into chunks or blocks each holding different parameter value. These block values are then transferred to different cloud servers to avoid data breaching or any other cyber-attacks. The storage of key values in different cloud server allows efficient data storage system. The experimentation shows us a clear difference on response time when the data size is increased. Moreover, the medical resource data can be well organized and stored properly in the defined method. Further researches are on progress for well-organized medical data storage and speed data access.

References

1. C. Konstantinou, M. Maniatakos, F. Saqib, S. Hu, J. Plusquellic, Y. Jin, Cyber-physical systems: a security perspective, in 2015 20th IEEE European Test Symposium (ETS), pp. 1–8 (2015)
2. V. Buzduga, D.M. Witters, J.P. Casamento, W. Kainz, Testing the immunity of active implantable medical devices to CW magnetic fields up to 1 MHz by an immersion method. IEEE Trans. Biomed. Eng. **54**(9), 1679–1686 (2007)
3. J. Zhang, N. Xue, X. Huang, A secure system for pervasive social network-based healthcare. IEEE Access **4**, 9239–9250 (2016)
4. Y.M. Huang, M.Y. Hsieh, H.C. Chao, S.H. Hung, J.H. Park, Pervasive, secure access to a hierarchical sensor-based healthcare monitoring architecture in wireless heterogeneous networks. IEEE J. Sel. Areas Commun. **27**(4), 400–411 (2009)
5. M.A. Khan, K. Salah, IoT security: review, blockchain solutions, and open challenges. Future Gener. Comput. Syst. (2017)
6. X. Yue, H. Wang, D. Jin, M. Li, W. Jiang, Healthcare data gateways: found healthcare intelligence on blockchain with novel privacy risk control. J. Med. Syst. **40**(10) (2016)
7. Z. Zheng, S. Xie, H. Dai, X. Chen, H. Wang, An overview of blockchain technology: architecture, consensus, and future trends, in *Proceedings—2017 IEEE 6th International Congress on Big Data, BigData Congress* (2017), pp. 557–564
8. M. Samaniego, R. Deters, blockchain as a service for IoT, in *Proceedings—2016 IEEE International Conference on Internet of Things; IEEE Green Computing and Communications; IEEE Cyber, Physical, and Social Computing; IEEE Smart Data, iThings-GreenCom-CPSCom-Smart Data 2016*, (2017), pp. 433–436

9. K. Christidis, M. Devetsikiotis, Blockchains smart contracts for the internet of things. IEEE Access **4**, 2292–2303 (2016)
10. N. Kshetri, Blockchain's roles in strengthening cybersecurity and protecting privacy. Telecomm. Policy **41**(10), 1027–1038 (2017)
11. A. Sharma, D. Bhuriya, U. Singh, Secure data transmission on MANET by hybrid cryptography technique, in *IEEE International Conference on Computer Communication and Control*, IC4 2015 (2016)
12. Y. Zhang, F. Patwa, R. Sandhu, Community-Based Secure Information and Resource Sharing in AWS Public Cloud, in *2015 IEEE Conference on Collaboration and Internet Computing (CIC)*, pp. 46–53 (2015)
13. M. Dark, Advancing cybersecurity education. IEEE Secur. Priv. **12**(6), 79–83 (2014)
14. W.J. Schünemann, M.O. Baumann, *Privacy, Data Protection and Cybersecurity in Europe* (2017)
15. N. Kshetri, India's cybersecurity landscape: the roles of the private sector and public-private partnership. IEEE Secur. Priv. **13**(3), 16–23 (2015)
16. M.R. Yuce, Implementation of wireless body area networks for healthcare systems. Sens. Actuat. A Phys. **162**(1), 116–129 (2010)
17. H.C. Keong, M.R. Yuce, Low data rate ultra wideband ECG monitoring system, in *2008 30th Annual International Conference of the IEEE Engineering in Medicine and Biology Society*, pp. 3413–3416 (2008)
18. M.R. Yuce, H.C. Keong, M.S. Chae, Wideband communication for implantable and wearable systems. IEEE Trans. Microw. Theor. Tech. **57**(10), 2597–2604 (2009)
19. J.Y. Khan, M.R. Yuce, F. Karami, Performance evaluation of a wireless body area sensor network for remote patient monitoring, in *2008 30th Annual International Conference IEEE Engineering Medical Biology Society* (2008)
20. J. Yusuf Khan, M.R. Yuce, G. Bulger, B. Harding, Wireless body area network (wban) design techniques and performance evaluation. J. Med. Syst. **36**(3), 1441–1457 (2012)
21. S. Al-Janabi, I. Al-Shourbaji, M. Shojafar, S. Shamshirband, Survey of main challenges (security and privacy) in wireless body area networks for healthcare applications. Egypt. Informat. J. (2016)
22. A.G. Fragopoulos, J. Gialelis, D. Serpanos, Imposing holistic privacy and data security on person centric eHealth monitoring infrastructures, *12th IEEE International Conference on e-Health Networking* (Healthcom, Application and Services, 2010), p. 2010
23. A. Papalambrou, A. Fragopoulos, D. Tsitsipis, J. Gialelis, D. Serpanos, S. Koubias, Communication security and privacy in pervasive user-centric e-health systems using digital rights management and side channel attacks defense mechanisms, in *2012 IEEE International Conference on Industrial Technology, ICIT 2012, Proceedings*, 2012, pp. 614–619
24. W. Badreddine, N. Khernane, M. Potop-Butucaru, C. Chaudet, Convergecast in wireless body area networks. Ad Hoc Netw. **66**, 40–51 (2017)
25. J.I. Naganawa, K. Wangchuk, M. Kim, T. Aoyagi, J.I. Takada, Simulation-based scenario-specific channel modeling for WBAN cooperative transmission schemes. IEEE J. Biomed. Heal. Informat. **19**(2), 559–570 (2015)
26. G. Anastasi, M. Conti, M. Di Francesco, A. Passarella, Energy conservation in wireless sensor networks: a survey. Ad Hoc Netw. **7**(3), 537–568 (2009)
27. G. Anastasi, M. Conti, M. Di Francesco, A. Passarella, An adaptive and low-latency power management protocol for wireless sensor networks, in *MobiWAC 2006—Proceedings of the 2006 ACM International Workshop on Mobility Management and Wireless Access*, vol. 2006, pp. 67–74 (2006)
28. S.H. Lee, J.H. Song, I.K. Kim, CDA generation and integration for health information exchange based on cloud computing system. IEEE Trans. Serv. Comput. **9**(2), 241–249 (2016)
29. F.B. Vernadat, Technical, semantic and organizational issues of enterprise interoperability and networking, in 2009 IFAC Proceedings Volumes (IFAC-PapersOnline), vol. 13, no. PART 1, pp. 728–733 (2009)

30. M.Z. Hasan, Intelligent healthcare computing and networking, in *2012 IEEE 14th International Conference on e-Health Networking, Applications and Services, Healthcom 2012*, pp. 481–485 (2012)
31. J. Walker, E. Pan, D. Johnston, J. Adler-Milstein, D.W. Bates, B. Middleton, The value of health care information exchange and interoperability. Health Aff. (Millwood), vol. Suppl Web (2005)
32. Y. Zhang, M. Qiu, C.W. Tsai, M.M. Hassan, A. Alamri, Health-CPS: healthcare cyber-physical system assisted by cloud and big data. IEEE Syst. J. **11**(1), 88–95 (2017)
33. J. Wan, H. Yan, H. Suo, F. Li, Advances in cyber-physical systems research. KSII Trans. Internet Informat. Syst. **5**(11), 1891–1908 (2011)
34. L. Zhou, V. Varadharajan, K. Gopinath, A secure role-based cloud storage system for encrypted patient-centric health records. Comput. J. **59**(11), 1593–1611 (2016)
35. B.J.S. Chee, F.J. Curtis, Cloud computing: technologies and strategies of the ubiquitous data center, in *Cloud Computing: Technologies and Strategies of the Ubiquitous Data Center*, pp. 67–90 (2010)
36. R. Sandhu, D. Ferraiolo, R. Kuhn, The NIST model for role-based access control, in *Proceedings of the Fifth ACM Workshop on Role-Based Access Control—RBAC'00*, pp. 47–63 (2000)
37. D.F. Ferraiolo, D.R. Kuhn, R. Chandramouli, Role-based access control. Components **2002**(10), 338 (2003)
38. K. Abouelmehdi, A. Beni-Hssane, H. Khaloufi, M. Saadi, Big data security and privacy in healthcare: a review. Proc. Comput. Sci. **113**, 73–80 (2017)
39. Y. Ashibani, Q.H. Mahmoud, Cyber physical systems security: analysis, challenges and solutions. Comput. Secur. **68**, 81–97 (2017)
40. H. Hu, Y. Wen, T.S. Chua, X. Li, Toward scalable systems for big data analytics: a technology tutorial. IEEE Access **2**, 652–687 (2014)
41. A.A. Cardenas, P.K. Manadhata, S.P. Rajan, Big data analytics for security. IEEE Secur. Priv. **11**(6), 74–76 (2013)
42. C. Tankard, Big data security. Netw. Secur. **2012**(7), 5–8 (2012)
43. H. Demirkan, A smart healthcare systems framework. IT Prof. **15**(5), 38–45 (2013)
44. G.E. Santagati, T. Melodia, An implantable low-power ultrasonic platform for the Internet of Medical Things, in *Proceedings—IEEE INFOCOM (2017)*
45. K. Bache, M. Lichman, *UCI Machine Learning Repository*. University of California Irvine School of Information, vol. 2008, no. 14/8. p. 0 (2013)

Chapter 4
An IoT Model to Improve Cognitive Skills of Student Learning Experience Using Neurosensors

Abhishek Padhi, M. Rajasekhara Babu, Bhasker Jha and Shrutisha Joshi

Abstract In a classroom, during the teaching period, there is a need of analyzing the basic level of understanding in a student in order to improve the teaching method for better teaching experience in a class. This model is required so that the concentration level of students can be monitored in a systematic manner, and after analyzing the concentration level, proper steps can be taken to improve it accordingly. This model presents designing an apparatus to record EEG waveform and then compare it to pre-recorded reading of different mind states using Arduino Brain Library and processing IDE to obtain the result as the emotion of the student. In the proposed method, EEG waveforms are obtained, which are the mathematical representation of the emotions; on analyzing those emotions, we can understand the level of concentration of the student in an efficient manner. It does not use any guesswork, and hence, the results obtained are reliable, and required actions can be taken on basis of that.

Keywords EEG · Waveform · Arduino Brain Library · Processing IDE
Electrodes · Neurosensor · IoT · Cognitive library

4.1 Introduction

4.1.1 Needs or Requirements

The education of students plays a vital role in the development of society, and so, student learning experience is a big area of interest. Proper learning or capturing of subjects taught in a class is majorly based on the states of brain and how students understand the concept taught. There are already a variety of methods which include interpreting the facial expression of the students, examining their hand–eye coordination and asking them question regarding class activities and studies. These methods have several drawbacks. They are inefficient as they involve very basic ways such as interpreting the facial expression/hand–eye coordination of the students and then changing teaching methods accordingly, but this method is not very reliable as they are very ambiguous and may lead one to take inappropriate actions to reach the goal [1].

This model concentrates on the brain activity which is the real factual data to analyze the state of mind to give appropriate result. The current state of brain or brain activity can be analyzed and studied with the help of brain waves since brain emits electrical waves which are called brain waves. These brain waves can be captured using EEG neurosensors, and they are referenced as EEG waveforms. The EEG signals are complex, multi-component periodic curves that are composed of high amplitudes which range between 1 and 50 Hz waves. These amplitude ranges are hence divided in eight parts, namely delta (1–3 Hz), theta (4–7 Hz), low alpha (8–9 Hz), high alpha (10–12 Hz), low beta (13–17 Hz), high beta (18–30 Hz), low gamma (31–40 Hz), high gamma (41–50 Hz). These states define the current state of mind such as relaxed, attentive, sleeping. These states can be recorded using EEG sensors, and hence, in this model, Mind Flex headset is used which contains NeuroSky chip which captures the brain waves and then converts these waves according to the frequency and amplitude to these eight parameters [2].

In the proposed model, Mind Flex headset is used with Arduino UNO as hardware to collect input from brain and get the data in the form of these parameters onto Arduino IDE. In Arduino IDE, this input will be analyzed and results will be extracted using the Arduino Brain Library which is a library, especially for brain waves analysis. Hence, these parameters will be analyzed and then processed with the help of processing IDE which will give the final result of how much a student has understood a particular topic.

4.1.2 Why This Work?

The ability to concentrate in class despite distraction, lack of interest or fatigue is an art that requires a lot of self-discipline and hard work. It is very difficult for one to focus on a specific task when there are multiple things going around, mind anyhow wanders away. Although the concentration time of a person and the factors that distract the person will vary from one person to another, hence, we can say that the actions that are required to improve the concentration level of a person will also vary and methodology will change for every other individual.

Talking about other feelings such as saying truth in difficult situations or expressing emotions, some people find themselves in a difficult situation where they deviate from the truth or they prefer to hide their emotions because it makes them nervous or they are just not capable of expressing their emotions even if they want to. These above-mentioned situations are required to be handled in an exclusive manner for every individual [3] (Fig. 4.1).

A. About the NeuroSky chip:

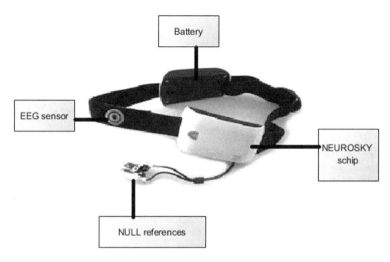

Fig. 4.1 Overview of NeuroSky headset

A few points of interest from the data given by the organization are as follows:

1. It is obtained from the product family, ThinkGear AM, where A corresponds to ASIC and M corresponds to module.
2. Next, demonstrate the number of the chip which is TGAM1, Revision Number 2.3.
3. So, the dimensions of the module are round about 29.9 mm × 15.2 mm × 2.5 mm (1.1 in. × 0.60 in. × 0.10 in.).
4. The module weight is 130 mg (0.0045 oz).
5. The working voltage of this module is about 2.97–3.63 V.
6. The maximum input noise which the module can possibly filter is 10 mV from peak to peak. We will then measure our noise and will ensure that the noise is in the module range for ideal outcomes.
7. Maximum power consumption of the module is 15 mA @3.3 V. We will check these quantities with a multimeter and will measure every parameter. It will be enjoyable to check these values by our own.
8. ESD protection of the gadget is 4 kV for the contact discharge and 8 kV for the air discharge. It is critical to take note of the fact that electrostatic discharge is the flow of electricity between two charged items caused by contact, dielectric breakdown, or electric shock. It is principally caused by a static charge of two bodies. The friction-based electricity can be built by induction or tribocharging (certain materials turn out to be electrically charged after they came into contact with various material) [4].
9. The gadget can communicate serially with 9600, 1200, 57,600 bps baud rate. There are arrangement pins by the assistance of which, we can change the baud rate.

10. Additionally, this TGAM1 chip can deal with just one EEG input, and we likewise need to process only one EEG channel, so it is able to utilize this chip [5].

4.1.3 ThinkGear Measurements (MindSet Pro/TGEM)

In this model, the Mindset Pro is put on forehead skin and ear and then the reference pickup potential or what is called voltage is found by the difference of dry sensor and the potential taken. The two are subtracted through basic rejection mode and served in as a one EEG channel and are amplified by 8000x to upgrade the faint/blackout quality of EEG signals. The obtained results or the signals are hence filtered by low- and high-pass analog-to-digital filter to retain signals in the range 1–50 Hz. Subsequent to correcting for possible aliasing, these signals are eventually sampled at 128 or 512 Hz.

The signal is analyzed every second in the given or available time space to identify and correct noise artifact, and at the same time retaining the original signals and hence, utilizing NeuroSky's restrictive calculations. A standard fast Fourier transform (FFT) is performed on the filtered signals; lastly, the signal is rechecked for any noise or artifacts in the frequency domain, by again utilizing NeuroSky's proprietary algorithms [5] (Fig. 4.2).

How does headset work? What does the ear movements signify? The working involves the following mentioned steps:

Step 1: The electrical impulses are sensed by the EEG sensor placed on forehead because of the neurons which are bombarded in the brain giving of the waves.

Step 2: The headset captures brainwave data, filtering out the environmental disturbances in the form of electrical noise, and interprets it with NeuroSky's attention and meditation algorithms.

Step 3: This mental state is then presented in the form of ear movements and shared.

From these ear movements, headset senses the attention and presents it in the form of ears shooting straight up. In relaxed phase, the ears droop down. Also, during highly focused and relaxed mode, the ears wiggle up and down [5].

In P3,
I. 1 is GND "−"
II. 2 is VCC "+"
III. 3 is RXD "R"
IV. 4 is TXD "T"
In P4,
I. 1 is VCC "+"
II. 2 is GND "−"

In P1,

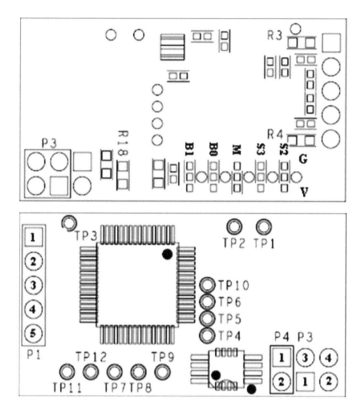

Fig. 4.2 Pin diagram of NeuroSky

I. 1 is EEG electrode "EEG"
II. 2 is EEG shield
III. 3 is ground electrode
IV. 4 is reference shield
V. 5 is reference electrode "RF"

4.2 Existing Methods

There are already a variety of methods in use. These methods include interpreting the facial expression of the students, examining their hand–eye coordination, reading their body behavior, and asking them question regarding what is going on in class. These methods are not much evolved as they involve very basic ways which are unreliable as they are very ambiguous and may lead one to take inappropriate actions to reach the deviated goal. Although this method cost nothing but is abysmally unreliable and totally ineffective as it involves guesswork and is not much backed by science [6–10].

In the paper titled "Cognitive neuroscience of creativity: EEG based approaches," Narayanan Srinivasan broadly contemplated the intellectual/cognitive neuroscience of imagination or creativity by utilizing the non-obtrusive electrical chronicles from the scalp called electroencephalograms (EEGs) and Event-Related-Potential (ERPs). This paper talks about significant parts of research utilizing EEG-/ERP-based examinations which includes chronicle of the signs, evacuating commotion, assessing ERP flags, and flag investigation for better comprehension of the neural associates of procedures engaged with innovativeness. Important factors are to be kept in mind while recording noiseless EEG signals. The recorded EEG signals have a possibility to be corrupted by various types of noise can be presented by following methods like the estimation of ERPs from the EEG signals by multiple trails. The EEG and ERP signals are additionally broke down utilizing different strategies including otherworldly investigation, rationality examination, and non-straight flag investigation. These investigation systems give an approach to comprehend the spatial actuations and worldly improvement of expansive-scale electrical movement in the cerebrum amid inventive undertakings. The utilization of this method will thus enhance our understanding of the neural and congnitive process. This also suggests methods for noise removal and follows the techniques like spectral analysis, coherence analysis, and nonlinear signal analysis [11].

In this paper titled "Evaluation of the NeuroSky Mindflex EEG headset brain waves data," J. Katona, I. Farkas, T. Ujbanyi, P. Dukan, and A. Kovari, informs that there is the difference in change in frequency which is observed in the spectrum of measured electric signals of the brain. The changes in the electric impulse that are generated during the various operations of the neurons are measured by the electroencephalograph (EEG) equipment. In this paper, the brain–computer interface [12] unit has been presented that is developed for further brain wave analysis and to ensure the detection of brain waves. This application can be used for acquiring the EEG data, processing, and visualization which could help in further researches in fields like medical research, multimedia applications, games. The novelty of this model is in measuring, collecting, processing and visualizing data using various software. The authors have made the program in such a way that it can be developed furthermore and new functions can be added due to its modular build with the development in the processing algorithms. Also, with this application, interfacing of the EEG headset device could be with other devices and the control of these devices can also be enhanced and solved, like the speed control and direction of a mobile robot. [2, 13].

In this paper "EEG-Related Changes in Cognitive Workload, Engagement and Distraction as Students Acquire Problem Solving Skills," Ronald H. Stevens, Trysha Galloway, and Chris Berka have begun to model changes in electroencephalography (EEG) to derive various measures of cognitive workload, involvement, and distraction as and when the individuals developed and improved their problem-solving skills. It was noticed that for the same problem-solving scenario(s), there were differences in the dynamics and levels of these three mentioned metrics. This paper found and observed that the workload was increased when students were assigned with problem sets of greater difficulties. A less expected outcome was, however, the finding that as skills increases, the level of workload did not decrease according to that. When these indices were measured and calculated across the navigation, decision, and display events within the simulation area, there were significant differences in workload and engagement observed. In the same way, event-related differences in these categories through a series of tasks were also often observed, but they highly varied across the individuals [14].

In this paper titled "Enhancement of Attention and Cognitive Skills using EEG based Neurofeedback Game," Kavitha P. Thomas and A. P. Vinod dealt with neuro-feedback, the self-regulation of brain signals recorded on utilizing electroencephalo-gram (EEG), which permits brain–computer interface (BCI) subjects to upgrade psy-chological and additionally engine which permits brain computer interface (BCI) user to improve their congnitive and motor function using various training meth-ods. Restorative impacts of neurofeedback (by the acceptance of neuroplasticity) on the treatment of individuals with neurological disorders, for example, dementia, attention-deficit hyperactive disorder (ADHD), and stroke have been accounted for in writing. In this paper, the authors research the effect of a neurofeedback based BCI game with respect to the improvement of psychological aspects of healthy subjects. Player's consideration-related EEG flag controls the BCI amusement. In the pro-posed preparing worldview, subjects play the neurofeedback amusement frequently for the duration of 5 days. Test investigation of player's consideration level (esti-mated by entropy estimations of their EEG) and the examination of intellectual test outcomes show the advantages of honing BCI-based neurofeedback diversion in the upgrade of consideration/psychological aptitudes. This paper examines the effect of the latest proposed neurofeedback game for improving the consentration abilities of the subject. The test demonstrate that the proposed neurofeedback paradigm moti-vates the player to enhance their entropy scores, improve attention level and thus achieve higher score in the game.

In this paper titled "A brain eeg classification system for the mild cognitive impairment analysis," A. Nancy, Dr. M. Balamurugan, and Vijaykumar S. observes that electroencephalogram (EEG) signals is a demanding and challenging task, and hence, some of the classification techniques which includes discrete wavelet trans-form (DWT), discrete cosine transform (DCT), and fast Fourier transform (FFT) are frequently used in the existing works across the world [29]. Yet, it had a few drawbacks; for example, the previously mentioned strategies speak to the structure estimations of input EEG signal in light of separated component of eye flickering estimation dataset [30]. To conquer this issue, this work proposed another framework: integrated pattern mining (IPM)—support vector machine (SVM) for the EEG signal order. In this work the EEG signals as input are pre proposed by using multiband spectral filtering and hence the specifications of the filtered signals are obtained. From that point onward, the ordinary or unusual mind states are grouped from the given flag utilizing SVM arrangement method. From the obtained output, the execu-tion of the proposed IPM-SVM technique is assessed and compare in terminologies like False Rejection Rate (FRR), False Acceptance Rate (FAR), Genuine Acceptance Rate (GAR), exactness, review, affectability, specificity and precision. The principle favorable position of this proposed framework is that it precisely characterizes the anomalous classification of intellectual weakness by enhancing the characterization execution of the signal classification framework [15].

In this paper titled "Discriminating different color from EEG signals using Interval-Type 2 fuzzy space classifier (a neuromarketing study on the effect of color to cognitive state)," Arnab Rakshit and Rimita Lahiri analyze that color perception is one of most important cognitive features in human brain and hence different cognitive activity is led by different color. Since color plays an important role, hence in this paper color-based recognition is shown using EEG sensors. Neuromarketing research based on color stimuli is a considerable tool for marketing research. It considered to consider first the color detection in mind in order to get different colors from EEG sensors. EEG sensors are hence used as a market based research tool in which the focus remains to detect various colors using the EEG sensors and thus the mentioned stimulus were obtained. This paper includes an interval type II fuzzy space classifier to differentiate between different stimuli which are considered for the ongoing experiment. Research says that red color has maximum classification rate and minimum is yellow. In this paper, red, yellow, blue, and green are the four colors to be considered for judgment. This uses the concept that human brain's color perception mainly occurs due to activation of lingual and fusiform gyri present in occipital lobes and left inferior temporal, left frontal, and left posterior parietal cortices where further information about the color is processed. EEG signals acquire the four different color stimuli, and Welch method is used for power spectral density estimation, and the extracted feature have been classified by IT2FS classifier. So this paper finally compares the results with other standard results and illustrates the activation of different brain regions by pictures.

In this paper titled as "Cognitive behavior classification from scalp EEG signals," Dino Dvorak, Andrea Shang, Samah Abdel-Baki, Wendy Suzuki, and André A. Fenton are discussing how EEG sensors are widely used these days and have power of accessing brain functions with extraordinary temporal resolution that is practically on the scale of milliseconds. Neurosensors are used in many fields now other than the psychiatric usage such as neurological, neurotherapy, medical, educational. In order to explore more about the potential of these sensors and to know what are the signals which are of interest for classifying diverse cognitive efforts, this paper has explored the details of how and why to use the EEG electrodes and what are the keys areas of signals and how they depict the different states of mind and how are the different signals depicted and in which form. This paper discovers the power of the EEG sensor electrodes by attaching it to the scalp and checking of the different types of signal with different frequency distribution and what level of area a particular frequency covers and hence in which areas the particular frequency range is used [16].

4.3   Proposed Method

This project involves the calculation of EEG wave on various aspects such as low alpha, high alpha, low beta, high beta, delta, theta, low gamma, high gamma, and then categorizing them according to their digital value. This digital value of the various aspects is then matched with the experimentally calculated value, and then, the level of emotions of the person is analyzed. By this way, accurate results can be known such as concentration level, level of distraction, attention level, and hence, proper steps can be taken to improve the cognitive learning of the student [17, 18].

Conclusion that can be included from the value of all these waves [19] is as follows:

Gamma Waves

If high: anxiety, stress, high arousal
If low: depression, ADHD, learning disabilities
Optimal: cognition, information processing, binding senses, learning, perception, REM sleep.

Beta Waves

If high: anxiety, high arousal, inability to relax, stress, adrenaline
If low: daydreaming, ADHD, depression, poor cognition
Optimal: memory, conscious focus, problem solving [20].

Alpha Waves

If high: inability to focus, daydreaming, too relaxed
If low: high stress, anxiety, insomnia, OCD [21]
Optimal: relaxation [20].

Theta Waves

If high: depression, ADHD, hyperactivity, impulsivity, inattentiveness
If low: poor emotional awareness, anxiety, stress
Optimal: creativity, emotional connection, intuition, relaxation.

Delta Waves

If high: learning problems, brain injuries, inability to think, severe ADHD
If low: inability to rejuvenate body, poor sleep, inability to revitalize the brain
Optimal: immune system, natural healing, restorative/deep sleep [19].

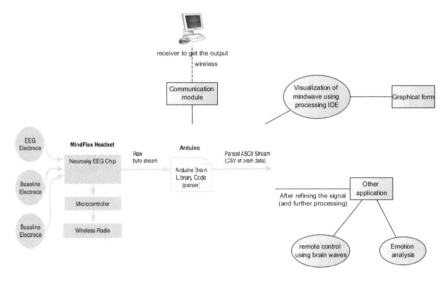

Fig. 4.3 Flow diagram representation of our model

The procedures followed while carrying out the work involve the various stages mentioned in the Fig. 4.3.

1. It starts with recording EEG waves using the NeuroSky chip. The analysis of the EEG waves is made using Arduino Brain Library. Next, various aspects of EEG waves are measured such as theta, delta, low alpha, high alpha, low beta, high beta, high gamma, low gamma.
2. Now, the visualization of the EEG waves is made using the processing IDE.
3. Next, the analyzed and recorded values are sent to other computers or databases.
4. After further analysis of various values, mind waves can be used to predict other emotions of an individual; it can also be used to control remote controls, etc. [22].

The implementation also involves the use of the Wi-Fi and Bluetooth modules to increase or widen the range/area and increase the reach of the model to far-off distances, including the places of the needy, so that it can help to uplift and develop them.

In Fig. 4.4, the four waveforms are of alpha, beta, theta, and delta, respectively. It is a sample which shows how the waveforms of various aspects of EEG waves look in a time interval of 1 s, when they are generated in a computer or any hardware.

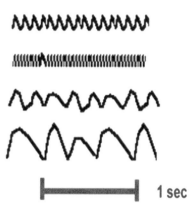

Fig. 4.4 Sample waveform representation of various aspects of EEG

1 sec

4.4 Result and Discussion

There is a fact related to Table 4.1 that higher value of low beta shows that the person has a higher level of concentration and attention [23]. So, from Table 4.1, we can conclude that concentration level is generally low in the morning (because of various reasons such as say feeling sleepy) which then gradually increases, and after a few hours, it decreases again (might be due to exhaustion) (Fig. 4.5).

A. Output Dataset

In Fig. 4.6, the representations are A1-signal strength, A2-attention, A3-meditation, A4-delta, A5-theta, A6-low alpha, A7-high alpha, A8-low beta, A9-high beta, A10-low gamma, A11-high gamma [24, 25].

Table 4.1 Variation of low beta value according to the class schedule

Class timing	Low beta value (10^{-5} Hz)
8:00 a.m.	27,971
9:00 a.m.	142,643
10:00 a.m.	74,039
11:00 a.m.	55,038
2:00 p.m.	30,022

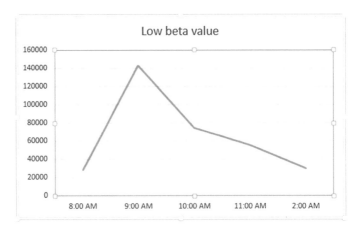

Fig. 4.5 Visualisation of Table 4.1

4.5   Conclusion

On feasibility background, currently, the headset is the most expensive part of our model because the headset has not officially been launched in India, so to buy the headset, one has to import it from America. But if it is made available in India, the expenses to make the model can reduce to its quarter price, which would make it very economical in comparison to heavy EEG machines available in hospitals. We found out that this model provides very accurate reading, and based on these readings, one can differentiate between different emotions of an individual. Its use can be expanded to various fields and occupations such as psychiatric, neurological, neurotherapy, medical, education [26]. So, with the inclusion of Brain Library in Arduino [27], the Mindflex can be used for various applications such as giving instruction and controlling hardware devices such as prosthetic arm [28], wheelchair. It does not need any prior knowledge/experience or any specialization to operate this device; hence, it can be used even by a layman. So, using the communication modules, live status of the mind of an individual can be analyzed from distance; hence, it can be used in long-distance learning. Using IoT with this module, its application can be expanded into vast areas and the data will not be bounded by the physical distance.

A1	A2	A3	A4	A5	A6	A7	A8	A9	A10	A11
200	0	0	319359	116282	19383	57371	27971	321320	309373	86135
200	0	0	301255	115732	47262	59332	74039	350273	270728	161825
200	0	0	1168533	106834	58588	64100	55038	173269	111704	46648
200	0	0	793149	1636301	168077	158127	142643	1209930	1230218	288919
200	0	0	954695	85642	10544	51821	30022	367489	288940	11735
200	0	0	544317	87652	42200	25115	45827	326977	320775	135755
200	0	0	935456	22587	7453	43027	23476	59545	25145	22275
200	0	0	1839457	184845	80700	51034	97170	119926	205145	49702
200	0	0	946367	55960	11626	27071	33248	334461	453190	86166
200	0	0	48410	34870	12468	14228	10865	155532	65640	21744
200	0	0	960394	274260	25617	27342	21752	44317	42477	20089
200	0	0	1781604	438267	51149	67627	86859	136935	153866	64013
200	0	0	1298747	106422	2310	39515	85292	314818	314021	135521
200	0	0	382878	168908	61401	27845	43644	444614	526489	11714
200	0	0	113324	168908	61401	27845	43644	444614	526489	117140
200	0	0	113324	34975	18440	11575	10405	99475	58875	26986
200	0	0	1496706	320525	18440	11575	10405	99475	58875	26986
200	0	0	1396670	66505	14359	48313	89609	318258	301285	186329
200	0	0	101334	72188	4964	8959	11544	166550	42783	27196
200	0	0	115178	34343	16823	13609	12242	98955	63517	28820
200	0	0	2956444	1527194	593580	326977	275357	710432	509659	339640
200	0	0	694938	82234	21663	29307	24727	162564	285513	257701
200	0	0	649938	82234	21663	29307	24727	157267	61048	31963
200	0	0	65064	56745	4143	8932	11868	157267	61048	31963
200	0	0	66006	26771	26686	20954	8160	99848	44890	17320
200	0	0	795319	556977	99890	33722	64960	251719	97599	34451
200	0	0	641952	249511	211873	96133	78760	323610	320621	115545
200	0	0	485139	182661	160866	130192	32696	322405	262371	170606
200	0	0	289023	305795	56532	40126	41681	533910	441419	12446
200	0	0	1168544	350004	52608	39834	53475	149289	207253	51418
200	0	0	1028779	222903	96140	63860	80282	144356	226090	124992
200	0	0	909833	819179	224653	112859	201830	881977	1347876	427584
200	0	0	107953	138598	17037	23159	21581	138450	76760	25300
200	0	0	556483	326598	93913	18808	34245	253530	269284	54436
200	0	0	1085446	209344	62056	67960	78063	158286	218878	36767
200	0	0	323816	154985	41497	82053	28027	295229	481335	106999

Fig. 4.6 Values of the various parameters of the EEG recording

References

1. K.V. Thomas, A.P. Vinod, A study on the impact of neurofeedback in EEG based attention-driven game (2016)
2. J. Katona, I. Farkas, T. Ujbanyi, P. Dukan, A. Kovari, Evaluation of the NeuroSky MindFlex EEG headset brain waves data (2014)
3. P. Sri Sai Chaitanya, S. Agnadi, A Review on improving technologies in wireless communications (2013)
4. M. Chaumon, D.V. Bishop, N.A. Busch, A practical guide to the selection of independent components of the electroencephalogram for artifact correction. J. Neurosci. Methods **250**, 47–63 (2015)
5. https://www.engineersgarage.com/articles/understanding-neurosky-eeg-chip-detail-part-213
6. P.S. Yalagi, T.S. Indi, M.A. Nirgude, Enhancing the cognitive level of novice learners using effective program writing skills (2016)

7. S. Ahmed, K. Li, Y. Li, H. Qureshi, S. Khan, Formulation of cognitive skills—a theoretical model based on psychological and neurosciences studies
8. S. Rabipour, A. Raz, Training the brain: fact and fad in cognitive and behavioral remediation. Brain Cogn. **79**, 159–179 (2012)
9. T.Y. Chuang, I.C. Lee, W.C. Chen, Use of digital console game for children with attention deficit hyperactivity disorder. US-China Educ. Rev. **7**, 99–105 (2010)
10. D. Plass-Oude Bos, B. Reuderink, B. Laar, H. Gurkok, C. Muhl, M. Poel, A. Nijholt, D. Heylen, Brain-computer interfacing and games, in *Brain-Computer Interfaces*, ed. by D.S. Tan, A. Nijholt (Springer, London, Chap. 10, 2010), pp. 149–178
11. N. Srinivasan, Cognitive neuroscience of creativity: EEG based approaches (2006)
12. A. Nijholt, University of Twente, Enschede, the Netherlands Imagineering Institute, Iskandar, Johor Bahru, Malaysia, The future of brain-computer interfacing
13. J.K. Nuamah, Y. Seong, S. Yi, Electroencephalography (EEG) classification of cognitive tasks based on task engagement index (2017)
14. R.H. Stevens, T. Galloway, C. Berka, EEG-related changes in cognitive workload, engagement and distraction as students acquire problem solving skills (2007)
15. A. Nancy, M. Balamurugan, S. Vijaykumar, A brain EEG classification system for the mild cognitive impairment analysis
16. D. Dvorak, A. Shang, S. Abdel-Baki, W. Suzuki, A.A. Fenton, Cognitive behavior classification from scalp EEG signals (2018)
17. R. Chai, Y. Tran, A. Craig, S.H. Ling, H. T. Nguyen, Enhancing accuracy of mental fatigue classification using advanced computational intelligence in an electroencephalography system, in *Proceedings of the 36th Annual IEEE International Conference of the Engineering in Medicine and Biology Society* (2014), pp. 1318–1341
18. http://dangerousprototypes.com/blog/2011/02/12/brain-wave-monitor-with-arduino-processing
19. http://neurosky.com/2015/05/greek-alphabet-soup-making-sense-of-eeg-bands
20. B.S. Zainuddin, Z. Hussain, Alpha and beta EEG brainwave signal classification technique: a conceptual study (2014)
21. https://researchpaper.essayempire.com/examples/psychology/ocd-research-paper
22. https://github.com/kitschpatrol/Brain
23. G. Bujdosó, O. Constantin Novac, T. Szimkovics, Developing cognitive processes for improving inventive thinking in system development using a collaborative virtual reality system (2017)
24. J. Kevric, A. Subasi, Comparison of signal decomposition methods in classification of EEG signals for motor-imagery BCI system. Biomed. Signal Process. Control **31**, 398–406 (2017)
25. C. Pedreira et al., Classification of EEG abnormalities in partial epilepsy with simultaneous EEG-fMRI recordings. Neuro-image **99**, 461–476 (2014)
26. A. Khong, L. Jiangnan, K.P. Thomas, A.P. Vinod, BCI based multi-player 3-D game control using EEG for enhancing attention and memory (2014)
27. http://www.edgefxkits.com/blog/arduino-technology-architecture-and-applications
28. https://www.engadget.com/2017/12/11/researchers-prosthetic-hand-lifelike-dexterity
29. K.P. Thomas, A.P. Vinod, Senior member IEEE and Cuntai Guan senior member IEEE, Enhancement of attention and cognitive skills using EEG based neurofeedback game
30. A. Rakshit, R. Lahiri, Discriminating different color from EEG signals using interval-type 2 fuzzy space classifier (a neuro-marketing study on the effect of color to cognitive state) (2016)

Chapter 5
AdaBoost with Feature Selection Using IoT to Bring the Paths for Somatic Mutations Evaluation in Cancer

Anuradha Chokka and K. Sandhya Rani

Abstract Nowadays, the research in bioinformatics helps in finding out numerous ways in storing, managing organic information, and developing and analyzing the computational tools for better understanding. So far, much of the research has been carried out to overcome the difficulties in experimental methods while storing vast amounts of the data in different sequencing projects. In this process, many of the computational methods and clustering algorithms were brought to light in the past to diminish blocks between newly sequenced gene and genotypes by applying identified jobs. The latest specific applications invented in bioinformatics are paving way for more advancement by adding developments in machine learning and data mining fields. Because of a large quantity of applications acquired by various feature encoding methods, the existing classification results remained inadequate. Hence, the present study is intended to create awareness among the readers on the various possibilities available in finding somatic mutations by using machine learning algorithm, AdaBoost with feature selection, a classification in various feature selection techniques with their applications, and detailed explanation on the distinct types of advanced bioinformatics applications. This study presents the statistical metric-based AdaBoost feature selection in detail and how it helps in decreasing the size of the selected feature vector, and it explains how the improvement can be attributed through some measurements using performance metrics: correctness, understanding, specificity, paths of mutations, etc. The present study suggests some IOT devices for early detection of breast cancer.

Keywords Bioinformatics · Somatic mutations · Machine learning · AdaBoost Feature selection · IoT

5.1 Introduction

It is found in previous investigations that tumor samples in cancer patients display several types of genetic defects which have been infected to the mankind during somatic mutation developments from a normal cell condition. Somatic mutations are

P. V. Krishna et al., *Internet of Things and Personalized Healthcare Systems*,
SpringerBriefs in Forensic and Medical Bioinformatics,
https://doi.org/10.1007/978-981-13-0866-6_5

accumulated in every cell continuously where the effect of one gene is dependent on the presence of one or more modifier genes. This phenomenon is known as epistasis which plays a vital role in molecular evaluation and while limiting the continuous flow of mutation built up. The size of epistasis connections relies on the fitness function of the space in all genotypes. Therefore, it can be stated that the genotypes noticed in the growth samples square measure the results of a varied set of alteration methods envisioning a posh fitness landscape. The somatic mutation forms helps to understand the progressive ways of the developments in cancer. This Significant information specifies how somatic mutations are influenced by the epistatic gene interactions among them. In this scenario, it is highly difficult to pull out how cancer is developing in the unknown fitness landscapes conditions, and examining such a huge data with hundreds of intermittently genes is one of the highly demanding areas in the research of bioinformatics. Since the existing computational methods are unable to overcome setbacks in the path of success, there is an urgent need to develop appropriate methods for the advancement in medicine. The AdaBoost algorithm is now a well-known and deeply studied method to build ensembles of classifiers with very good performance with high accuracy. This study also focuses on IoT a latest technology which is being adopted in the healthcare systems to detect and diagnose the cancers earlier to save the lives of the people as well as the money of the victims.

5.1.1 AdaBoost Technique

AdaBoost (adaptive boost) is a machine learning classification technique, which builds ensembles of classifiers in order to give good result [1]. This algorithm creates group of weak classifiers which form sequentially to give final classifier. Weight will be given to each set of training data, and the weight from weak classifier will be updated to the next classifier. This process will be continued until the last training data tests to get the final strong classifier. The weight for the weak classifier will be zeros likewise the accuracy [1]. When the weight increases, the accuracy for that classifier will also increase. Each training instance will be reweighted according to its misclassification by the previous classifiers.

5.1.2 Feature Selection Techniques

Here, we tend to report organic process progression methods for neoplasm samples from body part, brain tumor, respiratory organ, and female internal reproductive organ cancer problem persons (patients). EPPs area unit is derived for using a machine learning machine technique to reconstruct ancestral genotypes from observed growth genotypes, referred to as feature selection techniques (FSTs) [2]. The main purpose of the feature selections is manifold, and the very first crucial point is: (a) It is used to avoid overfitting, and it also improves the model performance in a great way, that is prediction of performance at intervals. The study of each case is supervised classifications and to have better even good cluster search or detection at intervals. The second point is (b) to have faster even cheaper models. The third point is (c) to be grateful for deeper underlying and neat processes which generates the data. As AdaBoost algorithm is having advantages over existing techniques, In this paper this algorithm is considered to develop a classification model. However, the benefits of FSTs are worth full. FS techniques dissent from or to each other at intervals of the approach, and this search is incorporated at intervals of the feature subsets. The FS techniques are broadly classified into three important categories; those models are filter ways, wrapper ways, and an embedded way. The filtering principles are used to assess how relevant the selected data at the intrinsic properties. In majority of cases, FS score is calculated and small scoring choices unit of measurement are removed. Afterward, this type of choices is taken as input to the classification formula for assessment. The Feature selection is a onetime process, and these can be used for development and analysis of different classifiers.[3]. That is, each feature is taken into consideration on an individual basis.Therefore, to address the ignoring the feature dependencies among the variables, filter principles were introduced. By rendering this method, the analysis of a specific set of choices is obtained by testing a specific classification model to a specific classification formula. The second method, wrapper ways utilises various searching algorithms to extract significant features. The third method of Feature selection techniques is termed as embedded technique.

5.1.3 Internet of Things (IoT)

Presently, the world is at the Internet of Things World Forum, we've been hearing a great deal about the transformational estimation of the Internet of things (IoT) crosswise over numerous enterprises—producing, transportation, horticulture, brilliant urban communities, retail, back, and medicinal services. Such a large number of new arrangements are in plain view that helps associations either spare or profit. In any case, in medicinal services, IoT can really accomplish more than that; it can possibly spare lives.

5.1.4 Challenges in Sequencing

Single cell sequencing (SCS) has so many recent and advanced methodologies which have come into picture to expose the growth of a tumor unsimilarity and well-endowed resolution at very high level. Even though there are multiple benefits in SCS, it has many of its own problems. The foremost problem is noise which is identified in different genotypes [4]. It is also observed in several instances that these genotypes include false +ve and false –ve mutations with missing values. Because of this persistent problem of noise, the clustering methods were unable to recognize the subpopulations in the sequenced cell and even a simpler task like mapping cells to clones has become a difficult issue to resolve. The second issue occurs in unnoticed subpopulations. Because of partiality in sampling, under sampling, or in the disappearances of these subdomains, the exemplificated cells are used to correspond to the division of the subdomains which emerges in the lumps total life history. Hence, approaches are required to understand the unnoticed ancestral subpopulations to find out the development of a tumor exactly.

5.2 Existing Models

Navodit Misra expressed that BML is a predicated model on a probabilistic biological process path from traditional genotype to other neoplasm genotype that incorporates a nonzero chance. The model BML initially estimates the chance that the selected combinations of mutations that reach extreme degree in each one cell population that is been evolved from a standard cell gene and can in the long run attain a neoplasm cell gene [5]. Here, these users can talk over with it called evolutionary genes G. The probability of these genes G, i.e., P(G) is the process of genes which makes equals the total of path chances for each mutation source from which the traditional genotype that it passes through the tip as a neoplasm genotype. Additionally, assume we tend to be had good information of the biological process ways followed by every neoplasm sample because it is evolved from the standard cell state. The BML prototype estimates that the mutation augmentation method is constant and consecutive, continuing one mutation at a time. BML estimates the biological process chances employing a graphical model referred to as a theorem network [6]. Theorem networks describe an outsized category of chance distributions which will be pictured as directed acyclic graphs (DAGs). They have been applied to organic phenomena analysis as well as copy variety variations in cancer. BML estimates ancestral genotypes by imputing possible biological process ways. The gathering of ways connecting a group of vertices to a typical vertex will invariably be pictured by the tree. As a result of not knowing the true paths followed by determined samples, we tend to perform an extra optimization step, we tend to perform an extra optimization step, wherever we tend to perturb the ways employing a category of tree rearrangements referred to as the nearest neighbor that is being interchanged and repeat this method till the formula encounters an area optimum in tree area. They restricted the BML prototype to co-occurrence the mutations which gives a reliable mark of positive hypostasis. BML will not be performing complete bootstrap analysis process for neoplasm-mutated genes which is applied on datasets on more recurrently antecedently possible. Edith M Ross identified an automatic technique called oncogenetic nested effects model (OncoNEM) which is used for reconstructing lineage clonal trees using somatic nucleotide types of multiple tumor cells that explains the structure of mutation framework of containing same objects of related cells. This method probably calculates genotyping errors and verifies unnoticed subpopulations [4]. It also calculates similar mutation framework of cluster cells into subpopulations [1]. This method is applied to two sets of information to verify the neoplasm cells on muscle-invasive bladder and neoplasm cells on vital thrombocythemia for identification of cancer cell on them.

5.3   Methodology

Many feature selection strategies are there in literature in order to perform dimensionality reduction for terribly huge data. Feature choice strategies provide North American country the simplest way of reducing computation time, up prediction performance, and a far better understanding of the information in machine learning or pattern recognition applications. In this paper, we offer a summary of a number of strategies gift in the literature. The target is to produce a generic introduction to variable elimination which might be applied to a good array of machine learning issues. We tend to concentrate on filter, wrapper, and embedded strategies. We tend to conjointly apply a number of the feature choice techniques on commonplace datasets to demonstrate the pertinence of feature selection techniques.

5.3.1   *Redundancy and Relevancy Analysis Approach*

Despite the spectacular achievements within the current field of feature choice, we have a tendency to observe nice challenges arising from domains admire genomic microarray analysis and text categorization wherever knowledge might contain tens of thousands of options. Initial of all, the character of high spatiality of knowledge will cause the questionable downside of curse of spatiality. Secondly, high-dimensional knowledge usually contains several redundant options. Each theoretical analysis and empirical proof show that besides impertinent options, redundant options additionally have an effect on the accuracy [7], speed, and vibrant of machine learning algorithms and sought to eliminate yet. Existing feature choice ways principally exploit two approaches: individual analysis and set analysis. In individual analysis rank options in keeping with their importance in differentiating instances of various categories and might solely take away impertinent options as redundant options doubtlessly have similar rankings. Ways of set analysis look for a minimum set of options that satisfies some goodness live and might take away impertinent options yet as redundant ones. However, among existing heuristic search methods for set analysis, even greedy sequent search that reduces the search house from $O(2N)$ to $O(N2)$ will become terribly inefficient for high-dimensional knowledge [3]. The restrictions of existing analysis clearly counsel that we should always pursue a special framework of feature choice that permits economical analysis of each feature connectedness and redundancy for high-dimensional knowledge.

5.3.2   Feature Redundancy and Feature Relevancy

In normal, feature selection has concentrated so far in studying the relevant features. Even though latest study has focused on the presence of feature redundancy along with its results, there is some work to be accomplished in the explicit treatment of feature redundancy [7]. With a view to achieve the target, this study presents a traditional method of feature relevance and also explains the reason why it is impossible to feature redundancy to deal with alone and also introduces a suitable formal definition for feature redundancy that leads to the removal of redundant features effectively. On the base of the definitions given by John, Kohavi, and Pfleger, the feature redundancies are divided into three categories. They are strong relevant features, weak relevant features, and irrelevant features. Let F be a full set of features, Fi a feature, and Sa_i $=$ Fa $-$ {Fa_i}. These three categories could be regularized in the following manner. Generally, these categories are in relation to feature correlation. It has been agreed that two features are redundant when their values are correlated fully (e.g., features F2 and F3). In practical situations, it is very difficult to fix feature redundancy where a feature is related to other sets [3]. Hence, we propose a feature redundancy to formulate a method to explicitly recognize and remove redundant features.

5.3.3   Defining a Framework of AdaBoost Technique with Feature Selection

To classify the given datasets accurately, we use the advance machine learning technique called AdaBoost (adaptive boost). It is machine learning's boosting technique which helps us to combine multiple weak classifiers into a final strong classifier [1]. To remove redundant features, the modern feature selection techniques should depend upon the method for the subset assessment that completely deals the feature redundancy with the support of feature relevance [2]. These modern techniques are able to show improvement in the results when we apply both combinations. However, the main drawback lies in this technique; unbearable computational cost in the search of subset made them weak while handling a huge amount of dimensional data. In view of finding out a suitable method for this issue, the study presents a new approach in AdaBoost with feature selection that completely overcomes the drawbacks in the previous methods by introducing an explicitly handling feature redundancy process. The main goal of present study is to find out somatic mutations and bringing differences between strong relevance and irrelevant redundancy. Identification of these differences can be achieved when the definition of relevance is completely understood and by the achievement of the following two steps [3]. First, we find out cancer mutations using AdaBoost technique by classifying given datasets. Second, by removing redundant features and subsets by considering relevant features of the relevance analysis the advantage of the modern process is dividing the redundancy and relevance in the analysis process. Hence, it can be understood that this method is an advanced and optimized technique when compared with previous techniques used. After performing number of random weak classifiers, the resultant sum of all the weak classifiers to have strong classifier of AdaBoost technique is (1).

$$H(X) = \text{sign} \left(\sum_{c=1}^{p} b_c H_c(X) \right) \tag{1}$$

Among nonlinear connection measures, several measures supported data concept of entropy, a life of the ambiguity of an uncertain changing variable. The entropy of the changing variable A is outlined as below (2).

$$E(A) = - \sum_i Pa(a_i) \log_2(Pa(a_i)), \tag{2}$$

The entropy changing variable of A after monitoring values of other changing variable B is defined below (3)

$$E(A/B) = - \sum_j Pa(b_j) \sum_i Pa(a_i|b_j) \log_2(Pa(a_i|b_j)), \tag{3}$$

The amount of the entropy that the changing variable of A decreases reflects extra info concerning A the changing variable provided by the changing variable B, and it is gain given by (4).

$$IG(A|B) = E(A) - E(A|B). \tag{4}$$

Since the data gain tends to the favor of options with additional values, it ought to be adapting with their correlating entropy. Therefore, we decide symmetrical ambiguity here below as (5).

$$US(A, B) = 2 \left[\frac{IG(A|B)}{E(A) + E(B)} \right] \tag{5}$$

5.3.4 Schematic Representation for the Proposed Algorithm

See Fig. 5.1.

5.3.5 Algorithm and Analysis

The function used in fourth line of the Algorithm 1 h $\rightarrow$ {0, 1} is described as h(a) = 1, when a $\geq$ 0, and h(a) = 0 when a < 0. The classifier Hp(xi) and yi represent the

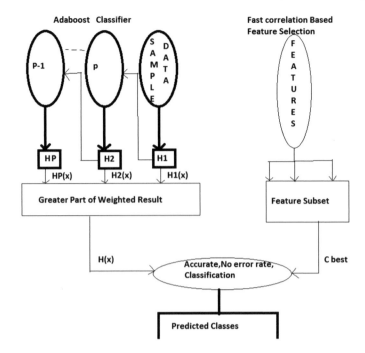

Fig. 5.1 Schematic representation for the algorithm AdaBoost with feature selection

values $\{-1, +1\}$, and the term errrate p is the weighted error rate. The final classifier is the summation of all the weak classifiers with sign [1]. Thereby, it finally classifies the result with great accuracy as mentioned in (1). The approximation methodology for connectedness associated redundancy analysis conferred before is completed by using an algorithmic specified by the authors in [2] (2) choosing predominant options from relevant ones. Using (1)–(4) for a knowledge set, it calculates the uncertainty of symmetrical (US) feature price for every feature.

5.3.6   IoT Wearables to Detect Cancer

The innovations in IOT related to wearable's, remote checking, execution helps to enhance well - being and well ness of the people and also to detect bosom disease. With inserted temperature sensors, this new sort of wearable innovation tracks changes in temperature in bosom tissue after some time. It utilizes machine learning and prescient investigation to recognize and group unusual examples that could show beginning period bosom growth.

A. **AdaBoost Classification Algorithm**.

Input: dataset $M = \{M_1, M_2, \ldots, M_N\}$ with $M_i = (x_i, y_i)$

where $x_i \in$ k and $y_i \in \{-1, +1\}$
P, the highest no. of classifiers
Result: A classifiers H: K $\rightarrow \{-1, +1\}$

1. Initialize the weights $W_i^{(1)} = \frac{1}{N}$, i $\in \{1, ..., N\}$, and set p=1; learner on **M** using
2. While p $\leq$ P do;
3. Run weak weights $W_i^{(p)}$ yielding classifier H_p: K $\rightarrow \{-1, +1\}$
4. Compute $errrate_p = \sum_{i=1}^{N} W_i^{(p)} h(-y_i; H_p(x_i))$;
5. Compute $b_p = \frac{1}{2} \log(\frac{i - errrate_p}{errrate_p})$ /* Weak learner **weight** */
6. For every sample i = 1, ..., L, update the weight
   $V_i^{(p)} = w_i^{(p)} \exp(-b_y H_p(x_i))$
7. Renormalize the weight: Calculate $S_p = \sum_{j=1}^{N} V_j$ and for i = 1, ..., N; $W_i^{p+1} = V_i^{(p)}/S_p$;
8. Increase the iteration counter: p $\leftarrow$ p++
9. End of while
10. H(X) $= sign(\sum_{c=1}^{p} b_c H_c(X))$

Algorithm 1. AdaBoost Structure Learning.

B. **Fast Correlation Feature Selection Algorithm**

Input: $C(f_1, f_2, ..., f_{N,d})$ /* **A** training Data Set * /
α,   /* predefined Threshold */
Output $= C_{best}$ /* Final Best Subset */

1. Begin
2. For i = 1 to N do begin
3. Calculate U $S_{i,d}$ for F_i
4. If(US$_{i,d} \geq \alpha$)
5. Append f_i to C_{list}^1
6. End;
7. Order C_{list}^1 in descending US$_{i,d}$ value
8. $F_v =$ get First Element C_{list}^1;
9. Do begin
10. $f_w =$ get Next Element (C_{list}^1, f_v)
11. if($f_w <>$ NULL)
12. do begin
13. $f_w^1 = f_w$
14. If (US$_{v,w} \geq$ US$_{w,d}$)
15. Remove f_w from C_{list}^1
16. $f_w =$ get Next element (C_{list}^1, f_w^1)
17. else $f_w =$ get Next Element (C_{list}^1, f_w)
18. End until ($f_w ==$ NULL)
19. $f_w =$ get next element (C_{list}^1, f_v);
20. end until ($f_v ==$ NULL);

Table 5.1 Feature set considered for fast correlation-based feature selection

Sl. no.	Features
1	Married status
2	Basis of diagnosis
3	Age
4	Occupation
5	Topography
6	Received surgery
7	Morphology
8	Received radiation
9	Stage
10	Survivability (classes)

21. $C_{best} = C_{list}^1$
22. end

Algorithm 2. Fast Correlation-Based Feature Selection Structure Learning.

The features considered for classifying cancer datasets here are shown in Table 5.1.

After applying fast correlation feature selection method as in the Algorithm 2, the subset of features from Table 5.1 features are age, occupation, and stage [8, 9].

The dataset was gathered from the databases of online open source as in Tables 5.2 and 5.3, Catalogue of Somatic Mutations in Cancer. The raw data comprises 49,875 patients' data, which is available in the database. Out of 49,875, 10,634 patients are having breast-related problems. By using AdaBoost technique, in primary-level classification we identified that 9426 patients are having carcinoma. In the second-level classification, it was observed that 1552 patients are having ductal carcinoma. Likewise, we can accurately classify the different sets of data. The classified result followed with the fast correlation-based feature selection is applied by considering the subset of features like age, occupation, and stage of cancer from Table 5.1 and able to predict the death or alive status of the patients as shown in Tables 5.2 and 5.3. As the objective of this paper is to predict survivability of the patients after the classification of somatic mutations detection. Using MATLAB tool, we were able to implement the code for the AdaBoost and feature selection technique and produced the results as in Figs. 5.2 and 5.3.

Table 5.2 Prediction of classes using AdaBoost with fast correlation feature selection algorithms for breast cancer datasets

Occupation		Stage		Prediction	
		2	3	Death	Alive
Workers	932	279	653	559	373
Managers	620	434	186	248	372

Table 5.3 Continuation of Table 5.2

Sl no.	Dataset	Samples	Carcinoma	Ductal carcinoma	Age		
					>22 <45	>45 <60	>60 <80
1	Brest cancer	10,634	9426	1552	456	894	202

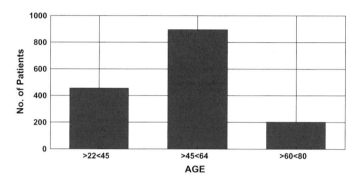

Fig. 5.2 Represents ratio of number of patients got cancer in a particular age

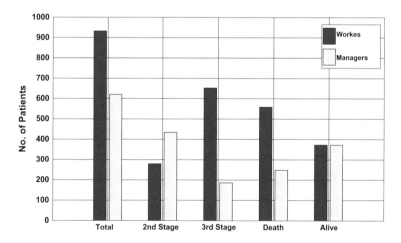

Fig. 5.3 Comparison chart between workers and managers

5.4   Conclusions

This paper discusses classification of cancer mutations using the fundamental details of AdaBoost algorithm and fast correlation feature selection technique. As a popular machine learning method, adaptive boost technique uses are in many different kinds of real-time applications, and feature selection's fast correlation method is helpful for finding the subset of features to reduce the redundancy; thereby combining both AdaBoosts with fast correlation technique, we get accurate classification of finding the somatic mutations in human beings. The requirements for the specific redundancy analysis are identified for tumor analysis in cancer and formulated a feasible definition to feature redundancy and selected features lineage, occupation, and stage of cancer to find mutations fast. By this, we were able to present a new type of approach which is suitable for the analysis of genes and related approach that practices the redundancy analysis. This feature rule is applied and examined in detailed experiments. The results of this AdaBoost with feature choice are confirmed by machine learning algorithms. Modeling the organic process events resulting in cancer and robustness scenery of cancer prisons guarantees the advanced applications in the analysis of cancer. Now, the AdaBoost with feature selection is a machine learning algorithm which permits the mutation of seemingly genotypes of ancestral and therefore the ways of reconstruction of accumulation in large detail accurately and quickly. Our methodology demonstrates its potency and effectiveness for feature choice in supervised learning in domains. Here, we spotlight the goal of AdaBoost by classifying mutations as parameters to drive analytical process. However the fast correlation recaptures the sequence of mutations mutated genes which recurrently accumulate. We have gathered datasets from online database sources. Our method is simple, fast, and accurate to know all tumor genotypes and able to impute evolutionary paths from the given sample datasets. It is very cheap in economy when compared with existing models. Therefore, the AdaBoost with feature selection is a special model having characteristics like correctness, specificity, understanding, and Darwinism of cancer genome. The best way to demonstrate the sickness potentiality specifies that 87% of probability to build up the illness is twofold mastectomy and, obviously, to have consistent mammograms and self-examination. Current techniques outside of the mastectomy are not generally ready to get growth early, that is IoT. Wearables, be that as it may, could give a contrasting option to the radical task. The American Cancer Society is foreseeing 1,658,370 new disease cases in 2015 and 589,430 malignancy passings in the USA. Incredible steps are occurring in the making of disease recognizing dress ready to identify bosom tumor in the beginning periods. On such item is the IoT inner wears. The gadget is a bra or other suit with implanted sensors that recognize little temperature changes in bosom tissue after some time. Enthused with the conceivable outcomes, Cisco supported a film in view of the innovation, called Detected.

References

1. A. Ferreira, Boosting algorithms: a review of methods, theory, and applications, in *Instituto de Telecomunicacoes* (Portugal, Chap. 3, April 2012), pp. 7–9
2. J. Thongkam, AdaBoost algorithm with random forests for predicting breast cancer survivability. IEEE Explore, 978-1-4244-1821-3/08 (2008)
3. L. Yu, Feature selection for high-dimensional data: a fast correlation-based filter solution, in *Proceedings of the Twentieth International Conference on Machine Learning* (Washington DC, 2003)

4. Y. Wang, X. Fan, A comparative study of improvements pre-filter methods bring on feature selection using microarray data. Health Inf. Sci. Syst. **2**, 7 (2014)
5. E.M. Ross, F. Markowetz, OncoNEM inferring tumor evolution from single-cell sequencing data. Ross and Markowetz Genome Biol. **17**(69) (2016)
6. N. Misra, E. Szczurek, Inferring the paths of somatic evolution in cancer, vol. 30 (Oxford University Press, May 2014), pp. 2456–2469
7. C.S. Attolini et al., A mathematical framework to determine the temporal sequence of somatic genetic events in cancer. Proc. Natl. Acad. Sci. U.S.A. **107**, 17604–17609 (2010)
8. Yu. Lei, Efficient feature selection via analysis of relevance and redundancy. J. Mach. Learn. Res. **5**, 1205–1224 (2004)
9. J. Thongkam, Breast cancer survivability via adaboost algorith ms: Australian computer society, in *Conferences in Research and Practice in Information Technology*, vol. 80 (2008)

Chapter 6
A Fuzzy-Based Expert System to Diagnose Alzheimer's Disease

R. M. Mallika, K. UshaRani and K. Hemalatha

Abstract Soft computing techniques came into reality to deal effectively with the emerging problems related to many fields. A medical diagnosis is totally based on human abilities, uncertain factors, ambiguous symptoms, high accuracy, and bulk of medical records. Soft computing techniques are suitable to obtain results in an efficient way in medical diagnosis. Fuzzy logic (FL) is one of the popular soft computing techniques. FL is a mathematical approach for computing and inferencing which generalizes crisp logic and sets theory employing the concept of fuzzy set. Fuzzy logic has been successfully applied in the fields of pattern recognition, image processing, knowledge engineering, medical diagnosis, control theory, etc., Alzheimer's disease (AD) is the most popular dementia in aged people. AD is an irreversible and progressive neurodegenerative disorder that slowly destroys memory, thinking skill, and degrades of the ability of performing daily tasks. Hippocampus is a key biomarker for AD to identify the disease at an early stage. To detect and diagnose Alzheimer's disease at an early stage, fuzzy logic is playing a vital role. In this study, computerized system for classification of AD was constructed using fuzzy logic approach, i.e., fuzzy inference system (FIS) to classify the subjects into AD, mild cognitive impairment (MCI), and normal control (non-AD) on the basis of visual features from hippocampus region.

Keywords Soft computing · Fuzzy logic · Medical diagnosis
Fuzzy inference system · Alzheimer's disease

6.1 Introduction

Soft computing consists of various computational techniques to study, model, and analyze complex real-time problems with imprecision, uncertainty, and partial truth [1]. Soft computing plays a vital role in the medical field. Soft computing techniques are effectively used to deal with the emerging problems related to medical diagnosis [2]. Alzheimer's disease is a type of dementia that leads to problems with memory, thinking, and behavior. It is an irreversible and progressive neurodegenerative

© The Author(s), under exclusive license to Springer Nature Singapore Pte Ltd. 2019   65
P. V. Krishna et al., *Internet of Things and Personalized Healthcare Systems*,
SpringerBriefs in Forensic and Medical Bioinformatics,
https://doi.org/10.1007/978-981-13-0866-6_6

disorder that slowly destroys memory, thinking skills, and degrades the ability to perform daily tasks [3]. Alzheimer's disease (AD) is the most popular dementia in elderly people worldwide [4]. AD diagnosis is based on the human abilities involving a number of uncertain and ambiguous factors. The vague and imprecision features of AD make the diagnosis process too complex and critical. To process the uncertainty and imprecision data involved in the symptoms of the Alzheimer's disease, [5]. Fuzzy logic technique is used in this study. FL resembles human decision making and has the capability to handle uncertainty, imprecision, and incomplete information. It has the ability to work from approximate reasoning and can provide ultimately a precise solution to the given problem [6].

In this study, brain MRI images from a publicly available OASIS database are considered [7]. The collected brain images are preprocessed and segmented to extract the hippocampus volume which is one of the key biomarkers of AD diagnosis [8, 9]. The main aim of this study is to develop an expert system with fuzzy inference system (FIS) to classify the brain MRI images into AD, mild cognitive impairment (MCI), and normal control (non-AD) subjects. Further, the performance of the proposed method is evaluated.

6.2 Literature Survey

Ali et al. [10] have presented various image mining techniques for the diagnosis of the Alzheimer's disease. Triangle area-based nearest neighbor (TANN), a new classification technique, is used. The performance of TANN using OASIS dataset is analyzed by comparing with k-nearest neighbor, support vector machine, decision tree, and Naive Bayes.

Keserwani et al. [11] proposed a methodology to classify the MRI images into AD disease or normal control based on Gabor texture features of hippocampus region. The performance of Gabor texture feature using OASIS dataset along axial, coronal, sagittal, and combination projections is evaluated by considering accuracy, sensitivity, and specificity.

Sampath and Saradha [12] extracted four different types of features from the functional magnetic resonance imaging (FMRI) AD images using Alzheimer's disease neuroimaging initiative (ADNI) database. The extracted first-order gray-level parameters, multi-scale features, textural measures, and moment invariant features are used for image segmentation. A novel method for AD image segmentation using self-organizing map network is proposed.

Madusanka and Cho [13] used multi-class support vector machine to classify the AD data using hippocampus morphological features.

Al-Naami et al. [14] have proposed a fusion method to distinguish between the normal and (AD) MRIs based on the use of low-pass morphological filters. Artificial neural network (ANN) is applied to test the performance of the fusion method.

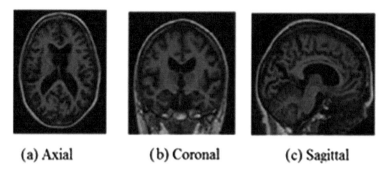

(a) Axial   (b) Coronal   (c) Sagittal

Fig. 6.1 Sample brain MRI images with three projections

6.3 Materials and Methods

6.3.1 Dataset

Data used in this study was obtained from the Open Access Series of Imaging Studies (OASIS) database. OASIS database consists of a cross-sectional collection of 416 T1-weighted brain magnetic resonance imaging (MRI) scanned images [15]. The each MRI image consists of three projections: axial, coronal, and sagittal. Sample MRI images of axial, coronal, and sagittal projections are presented in Fig. 6.1.

6.3.2 Proposed Methodology

A methodology is proposed for the detection of AD by extracting hippocampus of brain MRI-scanned images as hippocampus is a key biomarker for AD diagnosis. The proposed methodology consists of four phases: image preprocessing, segmentation, feature extraction, and classification. In first phase, MRI images are preprocessed to remove the noise and to enhance the image. In segmentation phase, the hippocampus region is segmented. In third phase, the hippocampus volume is extracted, and in final stage, the images are classified using fuzzy inference system (FIS). The four phases are shown in Fig. 6.2.

6.3.2.1 Image Preprocessing

To cope up noise and poor quality of medical images, preprocessing on MRI images is performed. MRI images of three projections, i.e., axial, coronal, and sagittal, are considered for preprocessing. At first, median filter is used to remove the noise which may be occurred due to light reflections. Later, the intensity level of images is

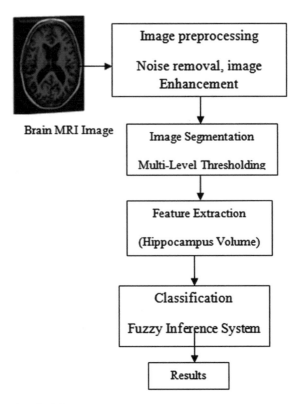

Brain MRI Image

Fig. 6.2 Proposed methodology

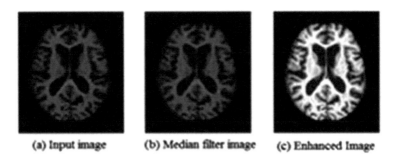

(a) Input image   (b) Median filter image   (c) Enhanced Image

Fig. 6.3 Image preprocessing results of MRI images in axial projection

adjusted by image enhancement to produce more suitable display for further image analysis. Figures 6.3, 6.4, and 6.5 present the results after preprocessing a sample image of axial, coronal, and sagittal projections, respectively.

   (a) Input image   (b) Median filter image   (c) Enhanced image

Fig. 6.4 Imagae preprocessing results of MRI images in coronal projection

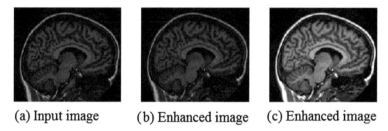

   (a) Input image   (b) Enhanced image   (c) Enhanced image

Fig. 6.5 Image preprocessing results of MRI images in sagittal projection

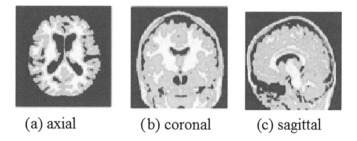

   (a) axial   (b) coronal   (c) sagittal

Fig. 6.6 Segmented AD MRI images

6.3.2.2 Image Segmentation

The resultant images after image preprocessing are segmented using multi-level thresholding method. After many pretests, multi-level thresholding method is considered for segmentation because it is able to segment the different regions of MRI images appropriately than other image segmentation methods. In particular, hippocampus region is segmented properly using multi-level thresholding [16]. Figure 6.6 presents the sample segmented MRI images of three projections.

6.3.2.3 Feature Extraction

In this phase, the features related to MRI images are obtained. Hippocampus volume is the feature extracted from the segmented images obtained from the previous phase. Hippocampus volume is a key biomarker sensitive to disease state and progression in Alzheimer's disease. From the segmented images, region of interest (ROI), i.e., hippocampus region, is selected manually to calculate the hippocampus area. The images in the dataset are of 1.2 mm thickness. Therefore, the same thickness is used to calculate hippocampus volume. The volume is calculated using the equation:

$$V = \text{Hippocampus area} * \text{Slice thickness}. \qquad (6.1)$$

6.3.2.4 Classification Using Fuzzy Inference System

Fuzzy inference system (FIS) comprises of fuzzy logic to accept the uncertain and imprecision data involved in the symptoms of the diseases and provides an exact output. It has the ability to work from approximate reasoning and can provide ultimately a precise solution to the given problem [17]. FIS consists of fuzzification, inference, and defuzzification. The internal structure of FIS [18] is presented in Fig. 6.7.

The process behind FIS involves certain stages:

- Fuzzification of the input variables.
- The fuzzy inference engine builds a decision-making logic based on the fuzzy rules.
- Defuzzification of the generated output to crisp values.

These steps involved in the process of developing FIS to classify the MRI images using extracted features are presented below.

Fuzzification and Determination of Fuzzy Rules
Hippocampus volume is used as the input to the FIS. In the literature [16], it is stated that the hippocampus volume of brain images will be in the range [3.623 3.873] cm^3

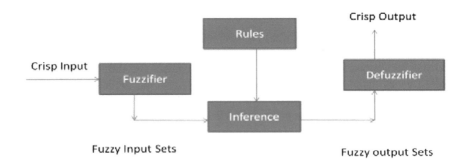

Fig. 6.7 Fuzzy internal structure

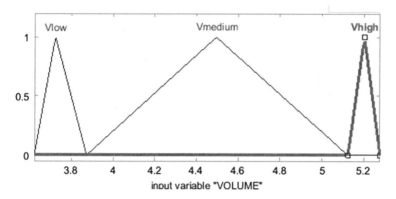

Fig. 6.8 Fuzzy MFs of hippocampus volume

for AD and for non-AD the hippocampus volume will be in the range [5.126 5.278] cm^3. Hence, universe of discourse (U) for V is determined as $U_V = [3.623\ 5.278]$, and it is portioned into three unequal length intervals. Each interval is represented by triangular membership function (MF) that corresponds to low, medium, and high hippocampus volume values. Based on the MF, the hippocampus volume is fuzzified as shown in Fig. 6.8. The interval ranges [3.623 3.873], [3.874 5.125], and [5.126 5.278] are considered for V_{low}, V_{medium}, V_{high}, respectively.

By examining the data, three fuzzy rules are generated. The generated rules are as follows:

- if hippocampus volume V is in interval V_{low}, then output is AD.
- if hippocampus volume V is in interval V_{medium}, then output is MCI.
- if hippocampus volume V is in interval V_{high}, then output is non-AD.

According to the membership degree of each variable and fuzzy rules, the output is generated. The generated output is decoded using defuzzification.

Classification
Brain images are classified using FIS, and the classification measures such as accuracy, sensitivity, and precision are calculated from confusion matrix. Confusion matrix is a table drawn between actual outputs and predicted outputs. It is used to measure the performance of a classifier. The equations of accuracy, sensitivity, and precision are given below

$$\text{Accuracy} = \frac{TP + TN}{TP + TN + FP + FN} \tag{6.2}$$

$$\text{Sensitivity} = \frac{TP}{TP + FN} \tag{6.3}$$

$$\text{Precision} = \frac{TP}{TP + FP} \qquad (6.4)$$

where TP = true positive, TN = true negative, FP = false positive, FN = false negative.

6.4   Experimental Results

Brain MRI images are preprocessed and segmented to extract the hippocampus volume. FIS classified the data of different projections into AD, MCI, and non-AD classes. The classification performance of FIS is evaluated using the measures presented in Table 6.1.

From the results, it can be observed that the classification measures of FIS for brain MRI images are high for axial projection with accuracy 86.53% and sensitivity 92.73% than other projections. The corresponding comparison is also depicted through bar chart in Fig. 6.9.

The obtained results are compared with the existing study Gabor filter-based classification (GFC) method conducted by Keserwani et al. [11] and tabulated in Table 6.2 for analysis.

Table 6.1 Classification measures of FIS

Brain MRI projection	Accuracy (%)	Sensitivity (%)	Precision (%)
Axial	86.53	92.73	91.71
Coronal	84.13	90.59	91.03
Sagittal	82.21	87.43	91.95

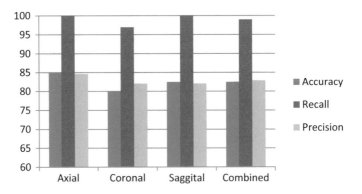

Fig. 6.9 Comparision of classification measures of FIS

Table 6.2 Comparison of the proposed method with other methods

Projection	Accuracy (%)		Sensitivity (%)	
	Proposed method	GFC method	Proposed method	GFC method
Axial	**86.53**	82.69	**92.73**	76.92
Coronal	84.13	72.92	90.59	73.07
Sagittal	82.21	76.92	87.43	80.76

From the tabulated values, it is clear that the proposed method accuracy is higher for all the three projections than the existing GFC method. Not only the accuray but also a significant improvement in the sensitivity of the proposed method is present which is desirable.

6.5 Conclusion

Nowadays, soft computing techniques are fetching importance to solve real-time problems, especially with complexity, imprecision, and uncertainty. In this study, brain MRI images from OASIS database are preprocessed and segmented. Hippocampus volume, one of the key biomarkers for AD diagnosis, is extracted from the segmented images. Fuzzy inference system (FIS) is used to classify the MRI images into AD, MCI, and non-AD subjects. The performance of FIS classifier is compared with existing works.

References

1. S. Rajasekaranand, G.A. VijayalakhsmiPai, *Neural Network, Fuzzy Logic, Genetic Algorithm-Synthesis and Application* (2011)
2. S.N. Sivanandam, S.N. Deepa, *Principles of Soft Computing* (Wiley, India)
3. Alzheimer disease fact sheet, Alzheimer disease education and retrieval center National Institute of health, NIH publication number: 11–6423 (2012)
4. P. Vemuri, C.R. Jack Jr., Role of structural MRI in Alzheimer's disease. Alzheimers Res. Ther. **2**, 23 (2010)
5. Association Association, Alzheimer's disease facts and figures. Alzheimer's and Dementia **10**(2) (2014)
6. K.M. Bataineh et al., A comparison study between various fuzzy clustering algorithms. Jordan J. Mech. Ind. Eng. **5**, 335–343 (2011)
7. OASIS brain Alzheimer dataset, www.oasis-brains.org/
8. N. Villain, B. Desgranges et al., Relationships between hippocampal atrophy, white matter disruption, and gray matter hypometabolism in Alzheimer's disease. J. Neurosci. **28**, 6174–6181 (2008)
9. F.M. Martin, *Fuzzy Logic—A practical Approach* (Neil, Ellen Thro, Academic, 1994)
10. E.M. Ali, A.F. Seddik, M.H. Haggag, Automatic detection and classification of Alzheimer's disease from MRI using TANN. Int. J. Comput. Appl. (0975 – 8887) **148**(9) (2016)

11. P. Keserwani et al., Classification of Alzheimer disease using gabor texture feature of hippocampus region. I. J. Image, Graph. Signal Proc. **6**, 13–20, Published Online June 2016 in MECS (2016), http://www.mecs-ress.org/, https://doi.org/10.5815/ijigsp.2016.06.02

12. R. Sampath, A. Saradha, Alzheimer's disease image segmentation with self-organizing map network. J. Softw. **10**(6) (2015)

13. N. Madusankaand, Y.Y. Cho, Hippocampus segmentation and classification in Alzheimer's disease and mild cognitive impairment applied on MR images. J. Korea Multimed. Soc. **20**(2), 205–215 (2017), https://doi.org/10.9717/kmms.2017.20.2.205

14. B. Al-Naami et al., Automated detection of Alzheimer disease using region growing technique and artificial neural network. World Acad. Sci. Eng. Technol. Int. J. Biomed. Biol. Eng. **7**(5) (2013)

15. Open Access Series of Imaging Studies available, http://www.oasis-rains.org/app/template/Index.vm

16. A. Vijayakumar, A. Vijayakumar, Comparision of hippocampal volume in dementia subtypes. ISRN Radiol. **5**, Article ID 174524 (2013), http://dx.doi.org/10.5402/2013/174524

17. I. Krashenyi et al., *Fuzzy Inference System for Alzheimer's Disease Diagnosis* (Current Alzheimer Research, 2015) [submitted for publication]

18. W. Siler, J.J. Buckley, *Fuzzy Expert System and Fuzzy Reasoning* (2005)

Chapter 7
Secured Architecture for Internet of Things-Enabled Personalized Healthcare Systems

Vikram Neerugatti and A. Rama Mohan Reddy

Abstract The Internet of Things (IoT) is the emerging area. This technology is made to connect any object around us to the Internet with the unique IP, and these connected objects can be communicated each other remotely as per the user's convenience. It has applications in all most all the fields like industries, factories, environment, agriculture, transport, education, healthcare, energy, and retail. IoT leads to the new technologies like big data and cyber-physical systems. Connecting any object, from anywhere at any time, is not simple. It has various challenges like discovery, scalability, software complexity, interoperability, fault tolerance, security, and privacy. One of the major challenges is security. Due to the weak links used to connect the things to the Internet leads to many security issues in different levels of the IoT. This paper presents the various security issues and novel security architecture for the IoT-enabled personalized healthcare systems.

Keywords IoT · Internet of Things · Security · Architecture · Healthcare

7.1 Introduction

Since a decade, Internet of Things is evolving. The enabling technologies of IoT are the cloud computing, wireless sensor networks, communication protocols, big data analytics, embedded systems, etc. Internet of Things is a truly ubiquitous computing, i.e., anywhere, anytime for every one computing said by Weiser [1]. The phrase "Internet of Things" is first coined by the Kevin Ashton in 1999 at MIT. Nowadays, the mobile phones and data rate become very cheaper, the wireless communication devices becoming smaller and cheaper and the processing capabilities is more. So the smart phone becomes the mediator for things, Internet, and people [2]. It has vast application domains like transportation and logistics, healthcare, smart environments, Personal, Social and Futuristic applications of IOT [3]. The major relevant scenarios for this domain are shown in Fig. 7.1.

© The Author(s), under exclusive license to Springer Nature Singapore Pte Ltd. 2019   75
P. V. Krishna et al., *Internet of Things and Personalized Healthcare Systems*,
SpringerBriefs in Forensic and Medical Bioinformatics,
https://doi.org/10.1007/978-981-13-0866-6_7

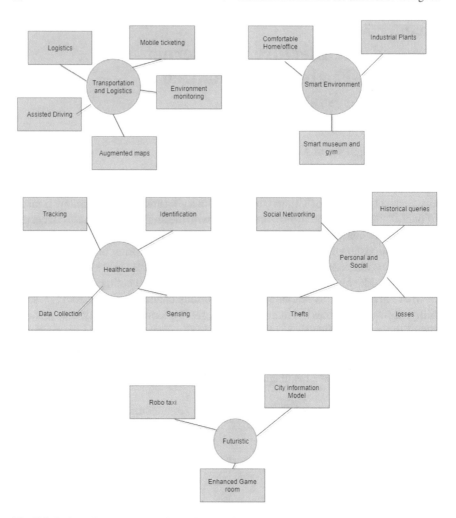

Fig. 7.1 Major relevant scenarios for various application domains in IoT

By 2025, the market share of the IoT applications is projected as healthcare 41%, manufacturing 33%, electricity 7%, urban infrastructure 4%, security 4%, agriculture 4%, resource extraction 4%, vehicles 2%, and retail 1% [4]. The projected analysis shows that major application of IoT healthcare leads a good market. Remote monitoring of the health plays a key role in the domain of the healthcare. If many IoT devices are connected to Internet, it may lead to many attacks. Particularly in the healthcare, if any data is misused, then the patient's life will be expired. The major attack in healthcare IoT is impersonate attack. In this attack, the intruder will pretend as he/she is a patient or doctor for collecting the data, etc. This leads to many security issues. It is necessary to develop security architecture in the area of healthcare IoT.

This paper proposed the novel security architecture for the IoT-enabled personalized healthcare systems.

The next sections in this paper are organized as follows: In Sect. 7.2, related works are discussed. The proposed architectures are discussed in the Sect. 7.3. Finally, in Sect. 7.4, conclusion and future work were discussed.

7.2 Related Work

In this research work we have [5] proposed an architecture for an IoT, it consists of the IoT nodes and the protocols where we can plant the sensors and the actuators and this architecture can be fit for different healthcare systems and various other applications in different disciplines. The experiments also proved that this is effective architecture for the IoT.

In [6], authors proposed a detection method for various attacks in the household appliance. The analysis was done in the theoretic fields by the simulation methods. In terms of the accuracy and localization, the proposed method proved as good. In paper [7], authors proposed a novel architecture for 5G smart dieses, in the layer-wise and comparison of the various diabetic versions and discussed about the personalized healthcare systems.

In the paper [8], proposed the various medical care services like the metabolic syndrome with the cloud-based personalized healthcare systems. Here service broker is responsible for the dynamic creations in the perspective of the users. This architecture can be used for the healthcare systems with the addition of some security features.

In paper [9] proposed the security features for the hybrid cloud architecture for the IoT with some security features. Particularly the authors had concentrated on the issues like scalability and interoperability, and some research challenges have been discussed.

7.3 Proposed Architecture

In the proposed system, the communication will be from the sensors that equipped with the patient's body to the cloud, and from the cloud to the corresponding doctor. The authorized doctor will access the data from cloud and treat the patient based on the obtained values. This will be done in layer-wise. The layers in IoT are user-side layer, edge-side layer, and cloud-side layer [10]. The security should be provided in and between those layers. The data flowed from one layer to other layer (Fig. 7.2).

Here the goal is to remote monitor the patient that living in their home from the hospital. To achieve this goal, the sensor values, communications links, data in cloud, etc., in all layers should be secured. To provide security for those layers in

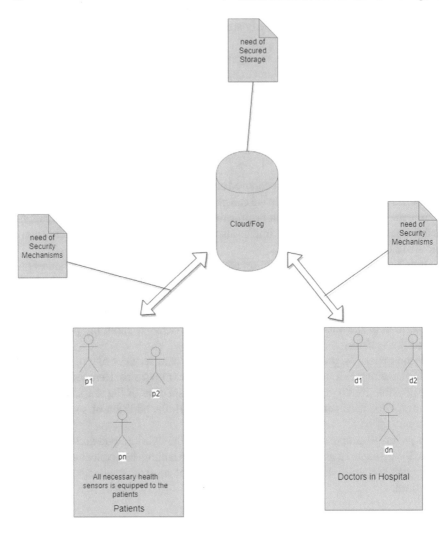

Fig. 7.2 Novel security architecture for the IoT-enabled personalized healthcare systems

IOT, first need to traced out the scope of attacks that occur in those layers and for those attacks the counter measures should be provided.

In the paper [11], authors proposed various attacks in the network layer of the IoT, particularly focused in the areas of network traffic, topology of network, and resources of network. Similarly, the attacks will be in the various layers of the IoT. The proposed architecture (Fig. 7.2) can mitigate this kind of the attacks (Fig. 7.3) in all layers.

The components in the architecture are patients, doctors, cloud/fog, security mechanisms, and secured cloud storage. In the patients component, various sensors will

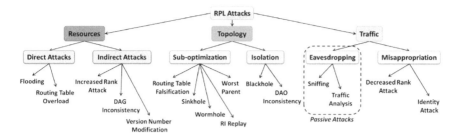

Fig. 7.3 Taxonomy of attacks against RPL networks [11]

be equipped in the patient's body and patient's home for continuous monitoring the patient and patient's health condition. The continuous data obtained by that equipped sensors will be stored in the component cloud. From that cloud component, the data can be accessed by the doctor module. The doctor's module consists of the doctors and there have a facility to collect the data of a patient from the cloud. In between all the three modules, doctor, patient, and cloud, secured communication links should be established, i.e., the security mechanisms module. For storing the data in cloud, the secured cloud storage component is used.

In the security mechanisms module, with the communication technology, there is necessity to integrate any one or more security mechanisms like the intrusion detection systems (IDS), authorization mechanisms, cryptographic mechanisms, and secured routing protocols. Simulataneously in the secured storage module need to use the secured data storage techniques like authorization and cryptographic algorithms.

7.4 Conclusion

The Internet of Things is the emerging technology, and it has various applications in all disciplines. This paper is focused on the personalized healthcare systems, and novel security architecture for the IoT-enabled personalized healthcare systems is proposed. This architecture leads to many research challenges like developing novel IDS, cryptographic mechanisms, and secured routing protocols. In the future work, we develop the novel intrusion detection systems for the RPL-based networks for mitigating the routing attacks like the black hole attack, worm hole attack, and rank attack.

References

1. M. Weiser, The computer for the 21st century. Sci. Am. **265**(9), 66–75 (1991)
2. F. Mattern, C. Floerkemeier, From the internet of computers to the Internet of Things (2011)

3. L. Atzori, A. Iera, G. Morabito, The Internet of Things: a survey. Elsevier Comput. Netw. **54**, 2787–2805 (2010)
4. A. Ala-Fuqaha et al., Internet of Things: a survey on enabling technologies, protocols, and applications. IEEE Commun. Surv. Tutor. **17**(4). Fourth Quarter (2015)
5. P.P. Pereira, J. Eliasson, R. Kyusakov, Enabling cloud-connectivity for mobile Internet of Things applications (IEEE, 2012)
6. J. Zheng, H. Qian, L. Wang, Defense technology of wormhole attacks based on node connectivity (IEEE, 2015)
7. M. Chen, J. Yang, J. Zhou, Y. Hao, J. Zhang, C.H. Youn, 5G-Smart Diabetes: toward personalized diabetes diagnosis with healthcare big data clouds. IEEE Commun. Mag. (2018)
8. S. Jeong, S. Kim, C.H. Youn, Y.W. Kim, A personalized healthcare system for chronic disease care in home-hospital cloud environments 978-1-4799-0698-7/13/$31.00 ©2013 IEEE
9. A. Sharm, T. Goyaly, E.S. Pilli, A.P. Mazumdar, M.C. Govil, R.C. Joshiz, A secure hybrid cloud enabled architecture for Internet of Things, 978-1-5090-0366-2/15/$31.00 ©2015 IEEE
10. V. Neerugatti, A.R.M. Reddy, An introduction, reference models, applications, open challenges in Internet of Things. Int. J. Modern Sci. Eng. Technol. (IJMSET) **4**(3), 8–15 (2017), ISSN 2349-3755
11. A. Mayzaud, R. Badonnel, I. Chrisment, A taxonomy of attacks in RPL-based Internet of Things. Int. J. Netw. Secur. **18**(3), 459–473 (2016)

Chapter 8
Role of Imaging Modality in Premature Detection of Bosom Irregularity

Modepalli Kavitha, P. Venkata Krishna and V. Saritha

Abstract Since last 60 years, bosom (breast) tumor is the major cause of death amid females worldwide. Earliest possible detection will raise the endurance rate of patients. Premature detection of bosom tumor is big challenge in medical science. Medical studies proven that imaging modalities like mammography, thermography, ultrasound, and magnetic resonance imaging (MRI) play a vigorous role to detect breast irregularity earliest. This paper enhances the knowledge on two imaging practices, one is mammography and another is thermography. It aids to identify the limitations in existing technologies and helps to plan the new methodology.

Keywords Bosom tumor · Mammography · Thermography · Ultrasound
MRI—magnetic resonance imaging

8.1 Introduction

The basic organic unit of humans is cell [1]. Generally, cells in the body are duplicated when new cells are needed but sometimes they duplicated out of control. Bosom tumor refers to the unrestrained duplicate of cells in particular site of body. The uncontrolled duplicate of cell may shape a swelling or microcalcifications or change the shape, which are rottenly called as tumors [2]. Bosom tumor is any form of malevolent tumor which develops from bosom cells [3]. It is most common cancer in all regions especially impedance rate is high in females [4].

According to World Health Organization's (WHO) Universal Agency for Research on Cancer, every year more than 400,000 women can expire due to breast cancer [5]. According to worldwide survey, in 2010 out of 23% breast cancer cases (more than 1.5 million patients) 14% death happens. In 2013, out of 232,340 new bosom tumor cases, around 39,620 patients were expired from bosom tumor in USA [1]. In 2014, the US cancer society estimated that there will be 232,670 new instances of intrusive bosom growth and 62,570 new instances of in situ bosom tumor [6]. As per the National Cancer Institute, 232,340 female bosom tumors and 2,240 male bosom growths are accounted for in the USA every year, and in addition around

© The Author(s), under exclusive license to Springer Nature Singapore Pte Ltd. 2019   81
P. V. Krishna et al., *Internet of Things and Personalized Healthcare Systems*,
SpringerBriefs in Forensic and Medical Bioinformatics,
https://doi.org/10.1007/978-981-13-0866-6_8

Table 8.1 A report on breast cancer risk versus age

Age	Risk
20–29	1 in 2000
30–39	1 in 229
40–49	1 in 68
50–59	1 in 37
60–69	1 in 26

39,620 passing's caused by the infection. In 2017, the American Cancer Society estimated that about 252,710 females will be determined to have intrusive bosom growth and about 40,610 will die from the disease.

The number of cases of Breast Cancer is more in developed countries while the incident rates are low. On the contrary, the incidence rates are high in developing nations while the number of cases currently is relatively low [7]. Breast cancer is increasing in India. There are an expected 1,00,000–1,25,000 new bosom tumor cases in India consistently. The quantity of bosom malignancy cases in India is evaluated to twofold by 2025. Premature recognition of bosom tumor can advance the chances of successful treatment and recovery.

Bosom malignancy is the second most driving reason for death in females from 15 to 54 years of age [8]. According to statistical reports, breast cancer risk increase depends on age [9]. Table 8.1 refers cancer risk rate depends on age.

Along with person's age, other factors include genetic factors, personal health history, diet, history of bosom tumor, having certain sorts of bosom protuberances, thick bosom tissue, estrogen introduction, stoutness, tallness, liquor utilization, radiation presentation, hormone substitution treatment (HRT), and certain employments all contribute to breast cancer risk. Most common symptoms of bosom tumor are bump or thickening in bosom (painless), irregular swelling of all or one a player in the bosom, constant skin bothering or dimpling, change in shading or appearance of areola, unusual discharge from the nipple other than breast milk.

Premature discovery is a viable approach to analyze and manage bosom tumor [1–5, 8, 9]. In this manner, precise analysis and effective balancing activity of chest tumor is an indispensable and basic issue in restorative science group. The bosom tumor takes 5 years' time to reach 1 mm size, an extra 2 years to achieve 5 mm and an additional 2 years to achieve 2 cm and so on [10]. This means if the number of years increases, the tumor size in bosom will increase, so premature detection and accurate treatment are mandatory to save the patient's life [11].

Imaging tools play a vigorous role in detection of bosom irregularities at premature phase. The bosom irregularities include denseness, tiny calcifications, size, shape variations, some visual signs such as skin refutation, the nipple discharge, skin thickness, and skin or nipple scratch [10]. Luckily, if bosom tumor is discovered precisely in a fundamental stage, confined tumors can be managed viably before the tumor spreads. Consequently, precise determination and compelling aversion of bosom tumor is a critical and enormous test in therapeutic science group.

The imaging modalities like mammography and thermography will help us to detect bosom irregularity at premature phase. Depending on the results, the patient will undergo the clinical inspections like ultrasound, MRI, biopsy to diagnose the bosom tumor. In this paper, I discuss existing system models mammography in Sect. 8.2, thermography in Sect. 8.3, and analysis in Sect. 8.4.

8.2 Mammography

Mammography is the present gold standard screening tool to detect bosom irregularity at premature phase [11–13]. Premature detection of bosom tumor will reduce the risk and increase the existence rate of patient. **Mammogram** is an X-ray image of bosom produced by FMRI or mammogram machines, whereas **mammography** is the procedure of translating the X-ray picture of bosom to discover the irregularities like tumor. It will help us to detect bosom irregularities like masses, microcalcifications, and architectural misrepresentations [14]. It is a low-measurement X-beam methodology that permits perception of the inside structure of the bosom [12].

There are three sorts of mammography: film, advanced (digital), and computerized bosom tomosynthesis. Film mammography utilizes broadly useful x-beam gear to record pictures of the bosom, while advanced mammography utilizes more specific automated hardware and conveys bring down dosages of radiation. Film mammography has been generally supplanted by advanced mammography, which gives off an impression of being much more precise for ladies more youthful than 50 years old and for those with thick (dense) bosom tissue [15–17]. **Analog mammogram is** the X-ray image of breast produced on film, and it is interpreted by radiologist manually, whereas **digital mammogram is** the X-ray image produced on computer screen, and it is interpreted by radiologist through CAD tools. Radiologist should be alert and must have proficiency to mark suspicious areas in film mammogram, whereas computer-aided detection tools support the radiologist to mark the doubtful areas in digital mammogram. If the radiologist does not have proper skill, then there may be a chance to get false-positives. Nowadays, the film mammography will be substituted by digital mammography because of its highly supported CAD tools. X-ray mammography does not give accurate results for the females having dense breasts [12].

In 2011, the FDA affirmed the utilization of advanced bosom tomosynthesis or three-dimensional (3-D) mammography, which builds a 3-D picture of the bosom with different high-determination x-beams, to be utilized as a part of blend with a 2-D computerized mammography picture [18]. The advantages and dangers of tomosynthesis in group hone are as yet being evaluated. A current report demonstrated the expansion of bosom tomosynthesis to advanced mammography may lessen false-positives and identify marginally more obtrusive growths contrasted with computerized mammography alone [19]. It may, when the 2-D pictures are created independently from the topographic pictures; ladies get about double the radiation measurements. As of late, the FDA affirmed the utilization of tomographic pictures to create engineered, customary 2-D pictures, consequently lessening the radiation measurements to that like traditional computerized mammography. This more up to date sort of mammographic screening is not yet accessible in all groups or completely secured by medical coverage.

Dense breasts are more probable to develop breast tumor, but mammography results are significantly less accurate in women with dense breast tissue [13]. The American malignancy society prescribes that patient ought to experience yearly screening start at the age 40 if they has family history of breast tumor [12]. But annual mammogram is not good for health, because the invasive ionic radiation may induce the cancer cells into our body or destroy the tissue through penetration [20]. Mammogram will catch susceptible state of bosom tumor 5 years early. It shows exact position of tumor, so it helps to give better treatment to the patient. Due to accessibility of convenient CAD tools, the mammography will become as an evergreen imaging tool to detect breast irregularity. Figure 8.1 demonstrates the mammogram attempted system of patients in laboratory.

The breast is pressed between flat plates during examination to get the clear images, so it causes pain and discomfort to the patient. Periodic mammogram is not possible to the middle-class people because of its high cost. X-ray mammograms are so noise; to reduce this, so many preprocessing algorithms are available. Due to ionic radiation, mammography is not effectual for females with thick breast, surgically altered breasts, pregnant women, women aged 40 and younger [21].

Mammogram Databases:

- LLNL/UCSF DATABASE
- NIJMEGEN DATABASE
- MIAS DATABASE
- American Cancer Society Surveillance Research
- National Cancer Institute SEER
- National Center for Health Statistics
- Mammographic Image Analysis Society (MIAS)
- www.stopbreasscancer.org
- www.medicalcconsumes.org
- www.cochrane.dk
- www.screening.dk

The mammographic picture experiences different stages, to recognize variations from the norm like preprocessing (clamor evacuation of a picture), discovering area of intrigue (differentiate upgrade, division, natives extraction, masses location), classification [10]. The immense change of biomedical fields gave hands-on procedures in picture-handling systems and influenced less demanding for the specialists to examine, to highlight extraction procedure, and to distinguish and order the tumors better [22].

Fig. 8.1 Mammogram test laboratory view

8.3 Thermography

Since 1960s, thermography is a screening tool to detect bosom temperature irregularity. It is an invasive, painless nature, radiation-free infrared imaging tool to detect the breast tumor at premature phase [21]. Premature discovery of bosom tumor will expand the endurance rate of patients. A thermogram is an infrared warm picture of bosom produced by FLIR E30 Infrared Camera [23]. We get JPEG format thermal images. Thermography is the process of interpreting the infrared image of breast to find the temperature irregularity. According to medical studies, the cancer will not attack both breasts at a time. That implies the temperature example of both left and right bosoms of a sound bosom thermogram is firmly symmetrical. Along these lines, a little irregularity in the heat pattern of left and right bosom may mean the bosom abnormality [21]. Modernized Infrared Thermal Imaging is an investigation, and it is delicate to physiological changes that are precancerous signs that may incite tumor. Infrared cameras can recognize emission in the infrared extent of the electromagnetic range and convey photos of that radiation, called thermograms [4]. Computer-aided diagnosis (CAD) tools will help the radiologist for automatic discovery of region of interest (ROI) [24]. Thermography method needs an infrared camera and PC to examine, distinguish, and create high-determination pictures of temperature changes in the bosom [4].

Thermography Principle:
"All articles more than zero Kelvin exudes infrared radiation. The Stefan–Boltzmann law gives the association between the infrared imperativeness and temperature. Infrared thermography suggests Emissivity of human skin is high (inside 1% of that of dim body) along these lines estimations of infrared radiation transmitted by skin can be direct changed over to temperature" [4, 23].

We have three ages of warm imagers. The first era warm imager (1970s) utilizes a small size linear array detector and two examining mirrors to get the picture. However, this imager gives poor clearness pictures. The second era warm imager (1980s) utilizes a small two-dimensional focal plane array instead of linear array, and still it uses scanning mirrors to generate picture. Still absence of lucidity is a major issue. The third era imager (1990s) having on-chip picture preparing ability and rather than mirrors it utilizes camera. So we get less commotion pictures [20]. Each of the three age imagers require a cooling framework support to see the warmth produced from the warmth source.

Presently, Now a days, uncooled cameras are available and they are generally utilized as a part of bosom imaging and programmed splendor alteration, smoothing, clamor expulsion is conceivable because of cutting edge computerized flag preparing [20].

Thermography is a 15-min test that helps the doctor to find the tumor at premature phase, but before attempting the test the chest regions are cooled with an aeration and cooling system and room temperature changed in accordance with 22 °C and obscured [4]. Fresh blood vessel development is important to manage the development of tumor [7]. The blood flow and metabolism are high in tumor so it will increase the skin temperature. Due to increased cell activity in tumor, the infrared radiations emitted from breast will grow up.

Temperature design appeared as hued picture, to distinguish the bosom anomaly where every shade adresses a specific extend of temperature. Shading designs are utilized to show signs of improvement come about clinically [4]. Each shade addresses a specific extent of temperature. The hued picture demonstrates asymmetry in bosom thermogram effectively.

Thermography reveals temperature irregularity in the form of image, and we know the body temperature is affected by different conditions. So we need to follow some recommended instructions mentioned below to reduce the error rate during thermographs [4, 23].

1. Don't expose the breast areas to the sun 5 days before the test

2. Don't use moisturizers, creams, powders, sprays on the day of examination of breast
3. Don't shave the breast area for 24 hours before the scan
4. Don't take any treatment related to physical fitness before 6 h of the scan
5. Don't undergo any other breast tests like ultrasound, mammography before 24 hours of the scan
6. Don't do bath before 4 h preceding the test
7. Take light food only
8. Don't take alcohol, smoke before 24 hours preceding the test
9. If any other health problems like fever, infections are there, then inform doctor prior to the scan
10. The chest is cooled 15 min before the scan with an air conditioner
11. The room temperature adjusted to 22° during the scan
12. Darkened the room during the scan to get better image.

Thermography is an instrument to gauge the temperature of a skin on a breast surface. Figure 8.2 demonstrates the thermogram attempted system of patients in laboratory. Dissimilar to mammography, the bosom parcel is confronted in front of the camera. It will detect irregularity of breast 10 years before the tumor formed. If the tissue is near to skin, it gives better results otherwise it does not gives good results. Thermography is a radioactivity-free, non-invasive, and inexpensive, patient-welcoming tool suitable for all ages of females; it works well for the female having dense breast, surgically altered breast, or even for pregnant ladies [21]. The small asymmetry between left and right bosoms in a thermogram refers susceptible state; to find this, there are so many methods available. One method exposes the bosom irregularity which is analyzing contralateral symmetry between breasts in a breast thermogram using statistical features like mean, entropy, standard deviation, skewness. To identify small asymmetry between breasts in a breast thermogram, we need to follow series of steps like preprocessing means removing unnecessary area, segmentation means extracting region of interest, feature extraction means identify the characteristics of an image, and feature analysis means evaluating those characteristics to reveal the normal or abnormal states [21, 23]. CAD tools are available to find region of interest automatically in a breast image. Another technology, dynamic area telethermography (DAT), is used to extract the diagnostic information from temporal changes in skin temperature [25].

Thermogram databases are as follows:

- Database for Mastology (DMR)
- Internet Resources
- Hospital Universitario Antonio Pedro (HUAP)
- The Thermogram Center, Inc.
- Research Laboratories
- Online databases of projects like PROENG

Advanced help is accessible to process the thermogramic pictures. To process a procured picture, different advances included like preprocessing (assessing sign to

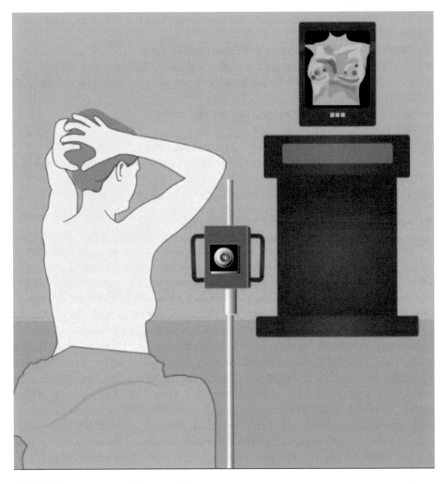

Fig. 8.2 Thermogram test laboratory view

clamor proportion), discovering locale of intrigue (bosom tissue partition), applying surface highlights like vitality, entropy, differentiate, contrast of variety, total distinction to discover the essentialness of asymmetry [26]. Many creators have proposed different strategies t fragment and identify hot areas like suspicious tissue locales in thermograms [27, 28].

8.4   Result Analysis

Imaging modalities play a vigorous role in fixing breast irregularity at premature phase. Premature phase detection of breast tumor will enhance the survival rate of

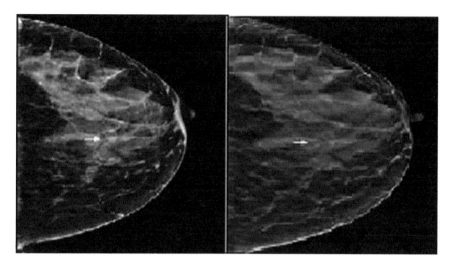

Fig. 8.3 Normal and abnormal mammograms

patients. So many imaging modalities available like mammography, thermography, ultrasound, and magnetic resonance imaging (MRI). In this mammography and thermography, both are complementary breast tumor risk assessment tools which one cannot replace the other.

Mammography is a low-measurement x-beam method that permits perception of inner structure of the bosom, and it is not suitable for younger women, pregnant women, surgically altered breast, who are on hormone replacement and not gives better results for the female having large or dense breasts. Doctors suggested female who are having family history of breast cancer must undergo annual mammogram after the age of forty but due to Ionic radiation exposure, it may not be preferable because it may destroy the tissue or may induce cancerous cells into our body. During the mammography, the breast is pressed between flat plates to get the quality images so it is painful, patient-uncomfortable high-noise costlier tool. Figure 8.3 shows normal and abnormal mammograms.

Thermography is a non-obtrusive and non-ionizing infrared imaging procedure that allows visualization of temperature pattern of the breast. It is radiation free so it applicable to all ages of females. According to medical studies, the tumor will not attack both breasts at a time and while forming the tumors, the skin temperature of breast will increase due to new blood vessel formation. So the small irregularity in left and right breasts' temperature pattern in a thermogram will result in susceptible state. It is a radiation-free, painless, patient friendly, inexpensive tool. Figure 8.4 shows normal and abnormal thermograms.

Mammography and thermography both are widely used to detect bosom abnormality at premature phase. Thermograms just recognize surface blood stream, so any disease development deeper may not be recognized except it occurs to be sufficiently substantial to bother the surface blood rivulet designs. Due to radiation,

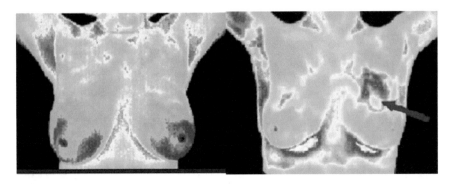

Fig. 8.4 Normal and abnormal thermograms

Table 8.2 Analysis of mammography and thermography

	Mammography	Thermography
Procedure	X-ray radiation	Infrared rays
Results	Shows internal structure of breasts like masses, microcalcifications, and architectural misrepresentations	Shows temperature pattern of breasts
Radiation	Ionic	Non-ionic
Detection	5 years early	10 years early
Dense breast	Not suitable	Suitable
Pregnant women	Not suitable	Suitable
Surgically altered breast	Not suitable	Suitable
Age	Best for the age above 40	Suitable for all ages
Digital support	Available	Available
Risk due to periodic undergo test	May be risk	Not a risk
Staff efficiency	Required	Required
Patient feel	Painful, uncomfortable	Painless, comfort
Cost	Expensive	Inexpensive

mammography will identify deeper calcifications also. Table 8.2 refers analysis of mammography and thermography.

8.5   Conclusion

Bosom tumor is the foremost cause of death among female worldwide. Premature detection of bosom tumor will increase the endurance rate of patient. Mammography and thermography both are widely used tools to detect bosom irregularity at premature phase, but premature detection of bosom tumor is acknowledged when different tests are utilized together. That means multimodal approach is preferable to get exact results in early phase. Multimodal approach incorporates bosom self-examination, physical bosom examination by specialist, mammography, thermography, ultrasound, MRI, and so forth.

8.6 Future Work

Premature detection of bosom tumor is a big challenge in medical community. Medical records say that premature detection of tumor will give 98% guarantee to save the patient. Currently, many imaging modalities like mammography, thermography, ultrasound, MRI are available to identify the bosom tumor. There is a need to develop advanced image processing tools.

Nowadays, Internet of things (IoT) plays vigorous role in medical field. So there is a challenge in our hands to develop an IoT tool to detect bosom irregularity.

References

1. C. Arya, R. Tiwari, Expert system for breast cancer diagnosis: a survey, in *2016 International Conference on Computer Communication and Informatics (ICCCI)* (IEEE, 2016)
2. K. Ganesan, et al., Computer-aided breast cancer detection using mammograms: a review. IEEE Rev. Biomed. Eng. **6**, 77–98 (2013)
3. M.F. Akay, Support vector machines combined with feature selection for breast cancer diagnosis. Expert Syst. Appl. **36**(2), 3240–3247 (2009)
4. S. Sheetal, A. Wakankar. Color analysis of thermograms for breast cancer detection, in *2015 International Conference on Industrial Instrumentation and Control (ICIC)* . (IEEE 2015)
5. T.M. Mejía et al., Automatic segmentation and analysis of thermograms usingtexture descriptors for breast cancer detection. *2015 Asia-Pacific Conference on Computer Aided System Engineering (APCASE)* .IEEE, 2015
6. X. Zhong et al., A system-theoretic approach to modeling and analysis of mammography testing process. IEEE Trans. Syst. Man Cybern. Syst. **46**(1), 126–138 (2016)
7. J.P.S. De Oliveira et al., Segmentation of infrared images: a new technology for early detection of breast diseases, in *2015 IEEE International Conference on Industrial Technology (ICIT)* (IEEE, 2015)
8. S.A. Taghanaki et al., Geometry based pectoral muscle segmentation from MLO mammogram views. IEEE Trans. Biomed. Eng. (2017)
9. D. Saslow et al., American cancer society guidelines for breast screening with MRI as an adjunct to mammography. CA Cancer J. Clin. **57**(2), 75–78 (2007)
10. B. Hela et al., Breast cancer detection: a review on mammograms analysis techniques, in *2013 10th International Multi-conference on Systems, Signals & Devices (SSD)* (IEEE, 2013)
11. K. Suresh, Labview implementation of identification of early signs of breast cancer, in *2014 International Conference on Electronics, Communication and Computational Engineering (ICECCE)* (IEEE, 2014)
12. M.S. Islam, N. Kaabouch, W.C. Hu, A survey of medical imaging techniques used for breast cancer detection, in *2013 IEEE International Conference on Electro/Information Technology (EIT)* (IEEE, 2013)

13. S.T. Luo, B.W. Cheng, Diagnosing breastmasses in digital mammography using feature selection and ensemble methods. J. Med. Syst. **36**(2), 569–577 (2012)
14. S. Paramkusham, K.M. Rao, B.P. Rao, Early stage detection of breast cancer using novel image processing techniques, Matlab and Labview implementation, in *2013 15th International Conference on Advanced Computing Technologies (ICACT)* (IEEE, 2013)
15. E.D. Pisano et al., Diagnostic accuracy of digital versus film mammography: exploratory analysis of selected population subgroups in DMIST. Radiology **246**(2), 376–383 (2008)
16. K. Kerlikowske et al., Comparative effectiveness of digital versus film-screen mammography in community practice in the United States a cohort study. Ann. Intern. Med. **155**(8), 493–502 (2011)
17. F.H. Souza et al., Is full-field digital mammography more accurate than screen-film mammography in overall population screening? A systematic review and meta-analysis. Breast **22**(3), 217–224 (2013)
18. US Food and Drug Administration, FDA approves first 3-D mammography imaging system (2016)
19. S.M. Friedewald et al., Breast cancer screening using tomosynthesis in combination with digital mammography. Jama **311**(24), 2499–2507 (2014)
20. H. Qi, N.A. Diakides, Thermal infrared imaging in early breast cancer detection-a survey of recent research, in *Proceedings of the 25th Annual International Conference of the IEEE Engineering in Medicine and Biology Society,* vol. 2 (IEEE, 2003)
21. U.R. Gogoi et al., Breast abnormality detection through statistical feature analysis using infrared thermograms, in *2015 International Symposium on Advanced Computing and Communication (ISACC)* (IEEE, 2015)
22. J. Kamalakannan et al., Identification of abnormility from digital mammogram to detect breast cancer, in *2015 International Conference on Circuit, Power and Computing Technologies (ICCPCT)* (IEEE, 2015)
23. A. Wakankar, G.R. Suresh, A. Ghugare, Automatic diagnosis of breast abnormality using digital IR camera, in *2014 International Conference on Electronic Systems, Signal Processing and Computing Technologies (ICESC)* (IEEE, 2014)
24. R.S. Marques et al., An approach for automatic segmentation of thermal imaging in Computer Aided Diagnosis. IEEE Latin Am. Trans. **14**(4), 1856–1865 (2016)
25. M. Anbar, C. Brown, L. Milescu, Objective detection of breast cancer by dynamic area telethermometry (DAT), in *Engineering in Medicine and Biology, 1999. 21st Annual Conference and the 1999 Annual Fall Meetring of the Biomedical Engineering Society BMES/EMBS Conference, 1999. Proceedings of the First Joint*, vol. 2. (IEEE, 1999)
26. S. Prabha, C.M. Sujatha, S. Ramakrishnan, Asymmetry analysis of breast thermograms using BM3D technique and statistical texture features, in *2014 International Conference on Informatics, Electronics & Vision (ICIEV)* (IEEE, 2014)
27. M. Etehadtavakol, E.Y. Ng, Breast thermography as a potential non-contact method in the early detection of cancer: a review. J. Mech. Med. Biol. **13**(02), 1330001 (2013)
28. T. Jakubowska et al., Thermal signatures for breast cancer screening comparative study, in *Engineering in Medicine and Biology Society, 2003. Proceedings of the 25th Annual International Conference of the IEEE*, vol. 2 (IEEE, 2003)

Chapter 9
Healthcare Application Development in Mobile and Cloud Environments

B. Mallikarjuna and D. Arun Kumar Reddy

Abstract The development of mobile healthcare application in pervasive devices has faced several challenges; common people can get benefited when healthcare services are developed in Android operating system. The development of mobile healthcare application on cloud computing environment deploys the data retrieval, availability of resources, security, and privacy. The aim of the proposed mobile healthcare application is to manage the healthcare information system among the various aspects. The healthcare application interacts with the cloud environment that facilitates to access the pool of resources and infrastructure on demand over the network. The interaction of cloud and healthcare application has been implemented using Amazon's S3 cloud service. The security of the system has maintained the concept of bring your own device (BYOD); we named the mobile health application as HealthKit, and it has been implemented in Android OS. The end of the HealthKit application is organized to manage the patient's health records; it provides the related healthcare information to physicians and nurses. The medical images are generated by the protocol digital imaging and communication medicine (DICOM) format; it was created by the Joint Photographic Experts Group committee (JPEG 2000) coding. The end of proposed HealthKit application has been tested and analyzed with the different network types.

Keywords Cloud computing · Amazon's S3 · Android operating system
DICOM format · JPEG 2000

9.1 Introduction

Mobile healthcare management focuses toward on e-health applications and retrieves or accesses the medical information anywhere and anytime through the mobile [1]. Mobile pervasive healthcare technologies can help both patient and physician it support of a wide range of healthcare services and provide various services such as mobile telemedicine, patient monitoring, location-based tracking medical services,

© The Author(s), under exclusive license to Springer Nature Singapore Pte Ltd. 2019   93
P. V. Krishna et al., *Internet of Things and Personalized Healthcare Systems*,
SpringerBriefs in Forensic and Medical Bioinformatics,
https://doi.org/10.1007/978-981-13-0866-6_9

emergency response, personalized monitoring, and pervasive access to the health information [1]. The health information management services in each hospital are different from each other; the services are used to share all information from every hospital to design a uniform information management system [1]. The existing system in healthcare management application has paper record information and prescription format to store the data and retrieve frantic process.

To shift traditional system to the e-healthcare system data accessing, data migration, storage, maintenance, update, etc., and to develop atomic distributed system for mobile application are achieved through the cloud computing is benefited to the common people. Cloud computing provides everything as a service in a ubiquitous and pervasive manner; it facilitates to access the shared resources and common infrastructure and offering the services on demand over the network. In this issue, we developed mobile healthcare application named as HealthKit, the application has been developed in Android operating system fully atomic distributed environment which is a pervasive healthcare information management system utilizing in cloud computing.

9.2 Related Work

The mobile healthcare information management services have been developed in iOS (iPhone operating system) for the latest version of iPhone 7 and 8 plus. There is no literature work utilizing cloud computing for developing pervasive healthcare management system by using Android OS. Several studies have been made in patient-related information, and in India and worldwide, many people are affected with deferent diseases; there is a need to develop mobile healthcare management services on Android OS for the benefit of common people [3–5]. The cloud computing is a paradigm which provides convenience to the users, the cloud computing provides on-demand network access to a shared pool of resources and configurable computing resources like networks, servers, storage, applications, and services are provisioned as metered on-demand service over the network that can be rapidly allocate the resources [6, 8].

The cloud computing model has various characteristics. One of the primary characteristic which involved in mobile health services is *on-demand self-service* it automatically consumer can access the network storage, server computing time [5, 6]. The next primary characteristic is *Broad Network access*; the smart phone can access the resources through the standard network mechanism on heterogeneous platforms [2].

9.3 Analysis of Health Diseases

There is a need for developing mobile healthcare application on Android operating system to bring into the common man. The healthcare statistical reports gathered and analyzed to plot the graphical reports. In India, eight states' statistical reports have been analyzed, adjusted of their ages greater than or equal to twenty which has been shown in Fig. 9.1. State-wise report of hypertension among tribal men and women their ages standardized prevalence of hypertension was men 27.1% and women 26.4% [3]. The observation has made highest hypertension noted in Odisha having 54.4–50% among men and women, the second one is Kerala 45–36.7%, and lowest state in Gujarat 11.5–07% [4].

In Fig. 9.2, it represents the country-wise death rates from chronic disease per every one lakh people; World Health Organization (WHO) released the statistical reports that the highest death rate was happened in Russian Federation where per every one lakh people 1000 members are died with chronic disease [4]. The lowest death rates are 220 noted in Canada, in India approximately 800 chronic deaths are observed in the year of 2005.

The WHO provides various statistical reports for the observation. In Fig. 9.3, thirty million people was died in 2005 from various disease such as heart disease, stroke, cancer, chronic diseases and diabetes [5]. 20% of these deaths was happened in high-income countries; high-income countries' death rates are less compared to low-income countries and 80% death rates was happened in low-income countries as shown in Figs. 9.3 and 9.4 [5].

In Fig. 9.4, it describes the death rates from low-income, middle-income, and high-income countries. Low-income countries' death rates are high especially among adults aged 30–69 years [5].

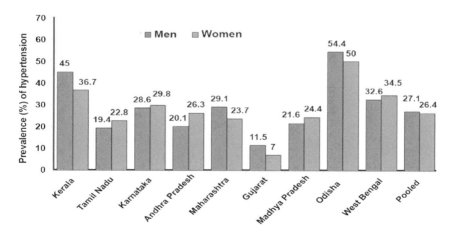

Fig. 9.1 State-wise report of hypertension among tribal men and women aged ≥ 20 years in state of India [3]

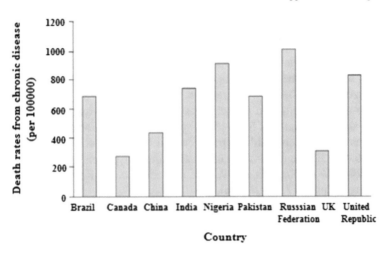

Fig. 9.2 Country-wise rate of death from chronic disease (per 100,000) [4]

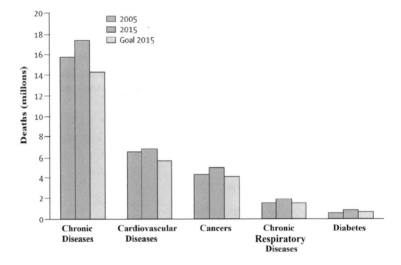

Fig. 9.3 Worldwide rate of death report from different diseases [5]

According to the statistical reports, high-income people death rates are less com-pared low-income people, high-income people having expensive treatment and their utilizing the expensive equipments like iPhone. The common man required mobile healthcare application on Android operating system; many mobile healthcare ser-vices are implemented in iPhone version 6 Plus–8 Plus which is iOS mobile operating

Fig. 9.4 Worldwide rate of deaths report from different diseases low-income, middle-income, and high-income countries [5]

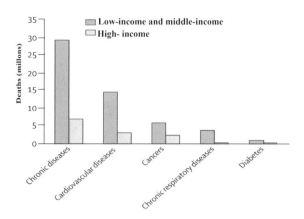

system. Those who are using iPhone, it can assist the people to save time managing their healthcare. The iPhone users are benefited to locate physicians nearer to them and manage their prescriptions or check their Health Savings Account (HSA) balance [5]. Hence, there is an enormous necessity to bring HealthKit application on Android OS.

9.4 Proposed Application Overview

Mobile healthcare application on Android OS provides the various services based on request from the user mobile. The set of cloud computing services are usually in the two platforms: front end and back end. The front interface provides the cloud platform that communicates directly with users mobile, and the back end allows the management of the storage content [1].

The healthcare organizations can assist the technical issues in bringing your own device (BYOD) implementation through a cost-effective way. The effective mobile device solutions utilize BYOD security and privacy issue [3]. The HealthKit application consists of various cloud services, the client running an application on Android OS, it consists of several modules such as patient health record, clinic module, patient module, cloud module and mobile user module.

Patient Health Record: In this module, the HealthKit application accesses and displays patient records that are stored into the cloud.

Medical Imaging Module: It displays the medical images of different types of images of protocol DICOM format, JPEG 2000 coding (Fig. 9.5).

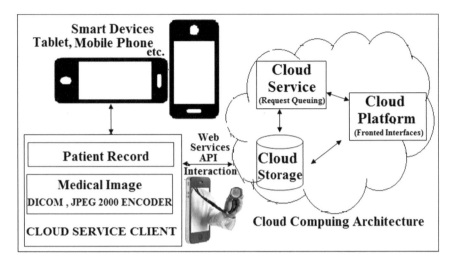

Fig. 9.5 Smart devices are connected to cloud computing architecture

In cloud computing architecture it provides specialized services such as cloud service, cloud platform and cloud storage etc., on mobile healthcare management systems. The mobile healthcare application enable simpler and quicker access to data such as patient records and medical images etc., [6]. The mobile healthcare application is a challenging task to develop on Android OS; it accesses the patient record, medical images, and cloud services to the client.

Clinic Module: This module works on routine activity of patients and physicians; it interacts with the web server to access the patient data, and it can access the clinic staff, the patients required for treatment and medicine required for people.

Patient Module: In this module, patients can get the medical information through their mobile phones; the patient authentication will be provided by the cloud server.

Cloud Module: In this module, cloud computing provides storage and services. The proposed healthcare application medical images are stored in Amazon S3 cloud service. The authenticated user accesses the features of client–server scheme in cloud computing paradigm.

Mobile User: The person can interact with the healthcare application by the client side; the client-side health application is a Web-based application standard port to be designed to connect port number 80 hosted on the Web-based application connect to the server side.

The healthcare application must have a network connection in order to be able to design HealthKit application for the following Algorithm 1.

Algorithm 1: *Design HealthKit Application on Android Operating System Overview*

Step 1: The Mobile User can retrieve the data, modify and upload
the medical content of patient medical images, health
records and bio medical signals etc. are needed to connect
cloud computing storage utilizing the web services.

Step 2:Query and view the patient health record management,
it provides the patient health record, related to bio signals,
image content

Step3: Display the medical images DICOM (Medical Image
Protocol Support) and it is incorporated with JPEG 2000 coding.

Step 4: At the end user can authenticated to cloud storage services.

As per the Algorithm 1, cloud computing paradigm, the healthcare application is designed and named it as HealthKit application.

9.5 Experimental Evaluation

The experimental evaluation has been conducted with T mobile phone; it contains Android OS version 8.0. The preparation of HealthKit application has been evaluated with different network types such as 3G, 4G, and WLAN shown in Table 9.1. The time is required to transmit data retrieval from the Amazon S3 Cloud service to HealthKit application which has been observed in Fig. 9.6. The existence results have been measured for transmission of medical images. There are number of medical images having different qualities such as occupational therapy (OT) which is 24-bit images, JPEG 2000 lossless color images, computed tomography (CT) uncompressed images, CT compressed JPEG 2000 images, medical representative (MR) JPEG lossless images, positron emission tomography (PET) JPEG 2000 lossy, and ultrasound sequence of 10 images, and JPEG 2000 lossless images having different file sizes and time intervals have been tested.

The effective results are carried out by the following graph as shown in Fig. 9.6.

Table 9.1 Medical image transmission using Amazon S3 cloud service

Image type (encoding)	File size (MB)	Time (size)		
		3G network	4G network	WLAN network
UT	6.8	42.532	20.865	7.894
CT (uncompressed)	0.528	4.023	2.035	2.382
CT (JPEG 2000)	0.102	1.223	0.985	0.892
MR	0.721	9.738	4.256	3.894
PET	0.037	0.923	0.159	0.793
Ultrasound	0.482	3.892	1.236	3.251

Fig. 9.6 Effective result on WLAN

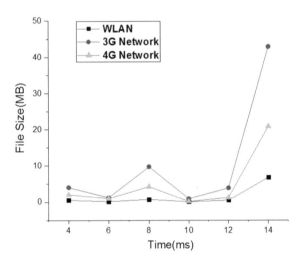

The experimental results show the WLAN having less time transmits the data in terms of response time in Amazon S3 cloud service. The HealthKit application works with the BYOD that provided by the IoT. The BYOD provides security and privacy to connect the various ubiquitous like tablets, mobile devices, smart phones, lap tops has become a major concern for healthcare organizations to access security and privacy through BYOD. The healthcare organizations focus on the concept of BYOD, it satisfies the patients, nurses and physician they are used portable device in their workplace through BYOD. In smart phone, Android operating system implemented the following HealthKit application; it can automatically detect the Bluetooth medical

Fig. 9.7 Evaluation of health checkup

devices and retrieve the data, and the data has been transferred to the Amazon S3 cloud service. The following Figs. 9.7, 9.8, and 9.9 are snapshots of the HealthKit mobile application while accessing through Amazon S3 cloud service DICOM header extraction JPEG 2000 progressive decoding a CT scan resolutions.

In Fig. 9.7, the evolution of health checkup in dashboard day-wise, week-wise, month-wise, and year-wise reports is displayed. The services of HealthKit application are atomic nature, that syncing to health application without the user having to do anything in a third-party app, making it easy for users to find health information in a single place.

In Fig. 9.8, feature plan needed to be take for health evaluation that will help common people to understand their health condition and where they would like to go, they can analyze his health condition and plan to get health check up. The HealthKit application can be designed to utilize low-order-income people; in Fig. 9.9, it describes the present-day health record, and it gives the present-day health record from wireless and networking to plan his/her weight, temperature, etc.

Fig. 9.8 Feature plan need to be take for the health evaluation

9.6   Conclusion

The present-day health communication mechanism is based on paper records, and prescriptions are old-fashioned, after health checkup to carry prescriptions which are inefficient and unreliable. In this contribution, mobile healthcare system is designed on Android OS in cloud computing environment. The proposed HealthKit application is implemented in atomic distributed manner, and the HealthKit application is bene-fited to common people, physicians, nurses, patients, and medical representatives as well as what medical devices have to use for patients.

In future, there is a scope to design and improve the services by advanced tech-niques on the mobile devices through voice recognition, exchanging images, or videos with new IoT techniques, and wide range of applications and service are most convenient to the user.

Fig. 9.9 Present-day health record

References

1. I. Maglogiannis, C. Doukas, G. Kormentzas, T. Pliakas, WaveletBased Compression With ROI Coding Support for Mobile Access to DICOM Images Over Heterogeneous Radio Networks. IEEE Trans. Inf Technol. Biomed. **13**(4), 458–466 (2009)
2. Kathleen Strong, Colin Mathers etc. Preventing cronic deseeses: how many lives can we serve? http://dx.doi.org/10.1016/S0140-6736(05)67341-2
3. R. George. Cloud Application Architectures: Building George Reese, Cloud Application Architectures: Building Applications and Infrastructure in the Cloud, O'Reilly Media, Paperback (April 17, 2009), ISBN 0596156367
4. I.I. Laxmaiah, Meshram etc. Socio-economic & demographic determinants of hypertension & knowledge, practices & risk behaviour of tribals in India. Division of Community Studies, National Institute of Nutrition (ICMR), Hyderabad, India, Indian J. Med. Res. **141**, 697–708 (2015)
5. WHO Library cataloguing-in-Publication Data Global status report on noncommunicable diseases 2010. ISBN 978 92 4 156422 9
6. The Android Mobile OS by GoogleTM. http://www.android.com/
7. The SQLite Database Engine. http://www.sqlite.org/
8. Amazons AWS Success Case Studies. http://aws.amazon.com/solutions/case-studiess

Chapter 10
A Computational Approach to Predict Diabetic Retinopathy Through Data Analytics

Ashraf Ali Shaik, Ch Prathima and Naresh Babu Muppalaneni

Abstract Making use of estimating methods in the field of medicine has been the powerful research recently. Diabetic retinopathy is a retinal disease which causes huge blindness. Recurrent screening for prior disease detection has been a highly labor force—and resource—powerful process. So computerized diagnosis of these diseases through estimating methods would be a great remedy. Through this paper, a novel estimation strategy for computerized disease prognosis is suggested, which utilizes retinal image analysis and mining methods to accurately differentiate between the retinal images as normal and affected. Eighteen feature relevance and three variations algorithms were analyzed and used to identify the contributing features that provided better conjecture results.

Keywords Diabetic retinopathy · Classification · CTree · SVMC
Bagging and boosting

10.1 Introduction

Exploration in neuro-scientific medicine shows that excessive pressure and glucose levels are a significant reason behind several critical health problems. Many of this astigmatism can lead to other problems in a variety of body parts. This paper gives awareness of Diabetic Retinopathy, being a classic disorder in the retina of the attention triggered due mainly to Diabetes [1] creating in decrease of vision and the second option being associated with a boost of pressure in the interest in the end creates harm to the optic nerve system [2]. Diabetic retinopathy is asymptomatic in your beginning stage and results say that treatment may be useful only once clinically diagnosed in the beginning. Standard screening process having risky of the condition might help discover the condition at the beginning stage. Finding retinal disorders in the characteristics made by screening process program is a moment task. Systematic diagnosis of the problem from the retinal data is an important part of recurring research [3–6].

In this research, we place attention to automated analysis medical prognosis of eyesight malocclusions (diabetic retinopathy) wherein the data is generally cleaned

© The Author(s), under exclusive license to Springer Nature Singapore Pte Ltd. 2019   105
P. V. Krishna et al., *Internet of Things and Personalized Healthcare Systems*,
SpringerBriefs in Forensic and Medical Bioinformatics,
https://doi.org/10.1007/978-981-13-0866-6_10

and statistical, GLCM organized, and bins setup measurements are computed. The classifier categorizes the choroid image to the problem category to which it belongs. This kind of research is geared toward robotic diagnosis of diabetic retinopathy through data and feature distinction using data gold exploration techniques.

Information mining maintains great prospective for the healthcare industry to allow health systems to systematically use information and statistics to identify lacking and best methods that improve care and reduce costs [7, 8]. Descriptive analytics describe what has happened. Predictive analytics [9] predicts what will happen. The disease estimations play a natural part in data gold mining. Info mining tools have recently been developed for effective examination of medical information, to be able to help clinicians for making better diagnosis to be treated purposes.

The two traditional algorithms, gold mining techniques proposed are CTree, support vector machine, also to classify the person with and without diabetes predicting the results Baye's Theorem is proposed.

(i) CTree (Conditional Inference Tree): It is the decision tree known as CTree, it is a non-variable class of r-Trees in tree-structured regression models into a well-learned theory of conditional procedures.

(ii) SVMC (Support Vector Machine Classifier): SVMCs also support vector classifiers [1] are supervised learning models with associated learning codes that analyze data used for classification and regression evaluation. Given a collection of training examples, each proclaimed as owned by one or the other of two categories, an SVMC training algorithm builds an unit that assigns new illustrations to one category or the other, rendering it a non-probabilistic binary linear class (although methods such as plot scaling exist to work with SVMC in a probabilistic classification setting). SVMC model is a representation of the examples as points in space, mapped so types of the independent categories are divided with a clear space that can be as extensive as it can be. New examples are then mapped into that same space and expected to participate in a category based on which aspect of the space they fall.

Furthermore, to performing linear category, SVMCs can proficiently do a nonlinear classification using precisely what is called the kernel trick, with ought a shadow of doubt maps their inputs into high-dimensional feature spaces.

Data is not tagged, checked learning is difficult, and an unsupervised learning approach is essential, which makes an effort to find natural clustering of the information to types and then map new data to these created groups. The clustering standards which gives an improvement to the support vector devices is referred to as support vector clustering [2] and is often [citation needed] used in professional applications either when data is not tagged or when only some data is defined as a preprocessing for a category pass.

Bayes Theorem: Bayes divisers are a family group of simple probabilistic classifiers based with strong (naive) independence presumptions of the features Bayes classifiers are highly ductile, requiring a quantity of parameters linear in the number of value predictors in a learning problem. Maximum-likelihood training can be achieved by accessing a closed-form expression, [1] which takes step-wise time, so expensive iterative approximation is used for most various divisors.

10.1.1 Steps in Algorithm

1. Each single data test is viewed by an n size vector, Y sama dengan (Y1, Y2, …, Yn), describing sizes made on quality from n values, correspondingly A1, A2, An.
2. Assume that there are n classes, C1, C2, …, Cn. Given an anonymous data test, Y. No classes designate, the classifier will predict that X is one of the classes getting the best posterior possibility, trained and only when

$$P(Ca|Y) > P(Cb|Y) \text{ for all those a} <= b <= n \text{ and } b! = a$$

Thus, we increase P(Ca|Y). The category Ca that P(Ca|Y) is strengthened is known as the utmost posteriori hypothesis. Simply B theorem

$$P(Ca|Y) = (P(Y|Ca)P(Ca))/P(Y)$$

3. P(Y) is regular for all those classes; only P(Y|Ca)P(Ca) must be maximized. When the probabilities are not known, then it is often guessed that the similarly is likely, i.e., P(C1)=P(C2) sama dengan…=P(Cn), and we would increase P(Y|Ca). Normally, P(Y|Ca) · P(Ca) is increased. Do not forget that the course probabilities may be approximated by P(Ca)=sa/s, where sa is the amount of training {samples of category Ca}, and s is the overall range of data training samples.

The manuscript is designed in this way. Section 10.2 targets the materials and methods. Section 10.3 shows the functionality measures while Sect. 10.4 presents tools used and experiment results. Section 10.5 concludes the research work.

10.2 Methodology

10.2.1 Description of Dataset

The dataset prevailed at https://archive.ics.uci.edu/ml/machine-learning-databases/00329/messidor_features.arff. This dataset contains 1151 records with nineteen features acquired from the Messidor image set to predict whether an image as indications of diabetic retinopathy or not.

10.2.2 Attribute Information

(0) Binary output of quality assessment is 0 = bad class 1 = sufficient class.

(1) Binary response to pre-screening, where one indicates severe retinal furor and 0 its shortage.

(2–7) Results of Messidor attribute detection. Each quality value is known for the number of Messidor attributes available at the confidence levels first = 0.5, …, 1, respectively.

(8–15) contain the same information as (2–7) for issues. However, as issues are represented by a group of points rather than the number of cells constructing the abrasion, these features are normalized by dividing the number of abrasion with the size of the ROI to compensate different image sizes.

(16) The Euclidean distance of the center of the macula and the middle of the optic disk to provide important information about the patient's condition. This feature is also normalized with the size of the ROI.

(17) Diameter of the optic disc.

(18) Binary reaction to the AM/FM-based classification.

(19) Class ingredients label: 1 = contains symptoms of DR (accumulative ingredients label for the Messidor classes 1, 2, 3), zero = no indications of DR.

Because Bayes algorithm will not grant regular data type, all the ideals in the dataset are cared for as categorical.

10.2.3 Cross-Validation

Cross-validation (CV) is the typical data retrieving way of examining performance of classification strategy. Mainly it is used to judge the challenge rate of an learning strategy. In CV, a dataset is portioned in n folds up, where each can be used for screening and the rest can provide for training. The process of screening and training is repeated n times so that all rupture of collapse can be used once for verification.

10.2.4 Classification Matrix

Classification matrix is a creation tool which is often used to provide the precision of the divisors in classification. It can be used showing the connections between final results and expected classes.

The entries in classification matrix have next meanings in framework of our research:

- p is the amount of accurate estimations value is negative,
- q is the amount of inaccurate estimations value is positive,
- r is the amount of inaccurate estimations value is negative,

- s is the amount of accurate estimations value is positive.

10.2.5 Bagging and Boosting

Bootstrap aggregating, also called bagging, is a machine learning ensemble meta-algorithm designed to increase the stableness and accuracy of machine learning algorithms used in statistical classification and regression. Additionally, it reduces variance and helps to avoid over-fitting.

Boosting is a machine learning ensemble meta-algorithm for mostly reducing tendency, and also variance in supervised learning, and a family of machine learning algorithms which convert fragile learners to strong ones. Algorithms that achieve speculation boosting quickly became simply known as "boosting."

Gradient boosting is a machine learning technique for regression and classification problems, which produces a prediction model by means of an ensemble of weak prediction models, typically decision trees.

10.3 Performance Measures

The concerned algorithm's ability to produce exact results was determined in this paper by the consumption of four characteristics: accuracy, sensitivity, specificity, and classification matrix.

10.3.1 Accuracy

In statistical scrutiny of binary classification, the F-measure is a way of computing a test's correctness. This considers both accuracy a and the recall r test to compute the report. a is number of correct excellent results divided by the amount of all excellent results, and r is amount of accurate excellent results divided by the amount of good results which should have recently been delivered.

10.3.2 Sensitivity

This is regarded as the opportunity that the relevant survey can be retrieved by the worried query. Also known as recall or true positive rate, it could be found in binary or nominal datasets but cannot be 100% depended after as a way of measuring developed.

10.3.3   Specificity

Additionally, it is called true negative rate and actions the percentage of negatives that are correctly thought as a result. Specificity belongs to the test's ability to effectively identify patients without a condition.

10.3.4   Classification Matrix

A classification matrix is a table that is often used to describe the performance of a classification model (or "classifier") on a group of test data for which the actual classes are known.

10.4   Tools Used and Results Discussion

The diabetic retinopathy dataset includes features extracted from the Messidor image set to predict whether an image contains signs of diabetic retinopathy or not. It includes 1151 instances with twenty attributes. In this dataset, all the attributes are numeric except the category label. The attributed used here are about quality assessment, Euclidean distance, size of the optic disk, and so on.

To obtain and calculate the test results of Bayes classifiers, the dataset instances are classified, i.e., size: 1151 is divided into training set (75%) = 863 and test set (25%) = 288.

To anticipate the occurrences of medical issues of individuals, it is extremely much necessary to examine the preceding data, utilizing data mining methods, especially for classification goal. As informed above, the classification methods that are taken into account are Bayes classifier, support vector machine classifier, decision tree, bagging, and boosting. In this paper, the research has been completed utilizing the wide open source data mining tool R.

The repository considered mostly includes 1151 instance of all the datasets and applied all the preferred classification algorithms as shown in Table 10.1. Out of this table, it could be recognized that the retinopathy expose higher accuracy of the diabetic dataset, it was examined in three specific ways: considering 1151 is divided into training set (75) = 863 and test set (25%) = 288 and last but not least with the help of nominal features. The results of the methods are shown in Table 10.2.

Finally, if we convert all the characteristics to nominal, the information becomes lossy which is not strongly recommended in the medical field as shown in Table 10.2. Because of this, it is extremely suggested maximum change of information into nominal should not exceed 50% of the complete data of the concerned attributes and this modification should be under the advice of doctors.

Table 10.1 Classification matrix

Actual	Predicted	
	−ve	+ve
−ve	p	q
+ve	r	s

Table 10.2 Comparison of results on the classification algorithms for diabetic retinopathy datasets over different instances and their data types. R was the tool used

Algorithm	Accuracy	Sensitivity	Specificity	Classification matrix		
Naïve Bayes	0.6146	0.5642	0.7714	*Prediction*	Reference	
					123	16
					95	54
Decision tree	0.6042	0.8633	0.3624	*Prediction*	Reference	
					120	95
					19	54
Support vector machines	0.5868	0.4029	0.7584	*Prediction*	Reference	
					56	36
					83	113
Bagging	0.6597	0.6835	0.6376	*Prediction*	Reference	
					95	54
					44	95
Boosting	0.5729	0.3669	0.7651	*Prediction*	Reference	
					51	35
					88	114

10.5 Conclusion

This kind of study evidently demonstrates the results are used for the information mining techniques of problem in medical directories.

In this paper, decision support machine classifier system was well suited for diabetic retinopathy. The machine can offer as training tool for medical students. Also, it can be heading be big hands for doctors. The machine can be further increased and extended; it can assimilate other medical properties besides in the Table 10.2 particular, and yes it can be integrated other gold mining techniques. Constant data can be utilized rather than just nominal data.

References

1. J. Han, M. Kamber, *Data Mining Concepts and Techniques* (Morgan Kaufman Publishers, 2006)
2. S.-C. Liao, I.-N. Lee, Appropriate medical data categorization for data mining techniques. Med. Inform. **27**(1), 59–67 (2002)
3. V. Balakrishnan, M.R. Shakouri, H. Hoodeh, L. Hakso-Soo, Predictions *Using Data Mining and Case-based Reasoning: A Case Study for Retinopathy*, vol. 63 (World Academy of Science and Technology, 2012)
4. R. Klein, B.E.K. Klein, S.E. Moss, T.Y. Wong, L. Hubbard, K.J. Cruickshanks, M. Palta (2004)
5. The relation of retinal vessel caliber to the incidence and progression of diabetic retinopathy: XIX: The wisconsin epidemiologic study of diabetic retinopathy, archives of ophthalmology, vol. 122, pp. 76–83
6. Ch-L. Chan, Y.Ch. Liu, Sh-H. Luo, Investigation of diabetic micro vascular complications using data mining techniques, in *International Joint Conference on Neural Networks (IJCNN 2008)* (2008)
7. X. Fang, Are you becoming a diabetic? A data mining approach, in *Sixth International Conference on Fuzzy Systems and Knowledge Discovery* (2009)
8. M. Shouman, T. Turner, R. Stocker, Using data mining techniques in heart disease diagnosis and treatment, in *Japan-Egypt Conference on Electronics, Communication and Computers* (2012)
9. M. Salehi, N.M. Parandeh, A. Soltain Sarvestani, A.A. Savafi, Predicted breast cancer survivability using data mining techniques, in *2nd International Conference on Software Technology and Engineering (ICSTE)* (2010)

Chapter 11
Diagnosis of Chest Diseases Using Artificial Neural Networks

Himaja Gadi, G. Lavanya Devi and N. Ramesh

Abstract The important health problems in the world are mainly caused due to chest diseases. A comparative chest disease diagnosis has been realized in this study by using the following neural networks such as multi-layer, probabilistic, learning vector quantization, constructive fuzzy, focused time delay, and generalized regression neural networks [1]. The back-propagation algorithm (It is a method used to calculate the error contribution of each neuron after a batch of data has been processed. It is the workhorse of learning in neural networks.) is the most popular algorithm in feed-forward neural network [1] with a multi-layer system. By adjusting the weights of artificial neural network while moving along the descending gradient direction is applicable to calculate the output error as well as the gradient of the error. The theme is to propose the implementation of back-propagation algorithm to compute and compare the percentage of the output accuracy, which is used for medical diagnosis on various chest diseases (i.e., asthma [2], tuberculosis [1, 3], lung cancer [4]; it is an iterative procedure that generates pneumonia).

11.1 Introduction

The increase in the count of people getting diagnosed with chest diseases every year in millions in the world, as the chest contains the main respiration and circulation organs which sustain some of the most critical life functions of the body. Back-propagation is the most popular learning technique in multi-layer networks. The ANN [1] is used to create network. Acute bronchitis,[1] ARDS[2] (Acute Respiratory Distress System), asbestosis,[3] and asthma which is a chronic disease which are occurred when the airway is swollen. The term acute is defined as a disease or condition characterized by the rapid onset of severe symptoms, and its death rate is approximately 5% of a people in a year. ARDS affects nearly 2 million people worldwide. The study aims

---

[1]It is the type of bronchitis that adds the cold and flu to an inflammation of the bronchial tube.

[2]It is occurred in the lungs when the oxygen level is low in the blood stream.

[3]Causes the lung tissues and the chest wall to get thicken and harden.

© The Author(s), under exclusive license to Springer Nature Singapore Pte Ltd. 2019   113
P. V. Krishna et al., *Internet of Things and Personalized Healthcare Systems*,
SpringerBriefs in Forensic and Medical Bioinformatics,
https://doi.org/10.1007/978-981-13-0866-6_11

to provide machine learning-based decision-support systems for contributing to the doctors in their decision of diagnosis.

11.2   Method

There are a total of 6 classes and 38 features that have been consisted in the considered dataset for chest disease measurement. The class distribution is done as follows:

- Tuberculosis
- COPD
- Pneumonia
- Asthma
- Lung cancer
- Normal.

All samples have thirty-eight features. These features are (laboratory examination): complaint of cough, body temperature, ache on chest, weakness, dyspnea on exertion, rattle in chest, pressure on chest, sputum, sound on respiratory tract, habit of cigarette, leukocyte (WBC), erythrocyte (RBC), trombosit (PLT), hematocrit (HCT), hemoglobin (HGB), albumin 2, alkaline phosphatase 2 L, alanine aminotransferase (ALT), amylase, aspartate aminotransferase (AST), bilirubin (total + direct), CK/Creatine kinase total, CK–MB, iron (SERUM), gamma–glutamil transferase (GGT), glucose, HDL cholesterol, calcium (CA), blood urea nitrogen (BUN), chlorine (CL), cholesterol, creatine, lactic dehydrogenase (LDH), potassium (K), sodium (NA), total protein, triglesid, and uric acid.

11.3   Neural Networks

An ANN is an information-processing paradigm that is inspired by the way human brain processes information. The original goal is to solve problems in the same way that a human brain would. The neural network [1] applications are pattern recognition, forecasting clustering data classification, healthcare problems [5–8]. ANN is applicable to analyze and make sense of the complex clinical data in medical applications, which is used as a tool to help doctors.

11.4   Types of Neural Networks

- Feed-forward neural network
- Feed-backward neural network (Fig. 11.1).

Fig. 11.1 Feed-forward
neural network

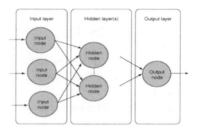

Feed-Forward Neural Network
This involves three different layers with feed-forward architecture. In this, the input layer contains the set of input neurons, which accept the elements of input feature vectors. The information is propagated layer by layer from input to output through one or more hidden layers.

Feed-Backward Neural network
The classification of the considered dataset is performed in the following neural networks:

- **Multi-layer neural network**: It is a more sophisticated artificial neural network, which is known as multi-layer neural networks because they have hidden layers.
- **Probabilistic neural network**: It is a feed-forward neural network, which is widely used in classification and pattern recognition problems. By this method, the probability of misclassification is minimized.
- **Learning vector quantization neural network**: It is a prototype-based supervised classification algorithm. This can be understood as a special case of an artificial neural network; more precisely, it applies a winner-take-all Hebbian learning-based approach.

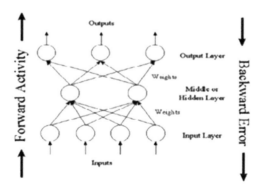

- **Constructive fuzzy neural network**: The constructive fuzzy neural network is used to classify the samples according to numerical attributes. This not only increases classification accuracy but also speeds up classification process.

- **Focused time-delay neural network**: These neural networks operate with multiple interconnected layers composed of clusters.
- **Radial basis function**: These networks have the advantage of avoiding local maxima in the same way as multi-layer perceptrons. These have been applied as a replacement for the sigmoidal hidden layer transfer characteristic in multi-layer perceptrons.
- **Generalized regression neural network**: It is a variation to radial basis neural networks. These can be used for regression, prediction, and classification. The advantage of generalized regression neural network is that it can handle noises in the inputs [9].

11.5   Back-Propagation Algorithm

The ANN [1, 10–12] has been trained by exposing it to sets of existing data where the outcome is known. A multi-layer network uses a variety of learning techniques; the most popular is back-propagation algorithm.[4] Learning is also called "training" in artificial neural networks [1] because the learning is achieved by adjusting the connection weights in ANN iteratively so that it gets trained. The back-propagation (BP) algorithm [13] is widely recognized as a powerful tool for training of the MLNN structures. However, BP algorithm suffers from a slow convergence rate and often yields suboptimal solutions [1, 14]. A variety of related algorithms have been introduced to address that problem, and a number of researchers have carried out comparative studies of MLNN training algorithms. The number of iterations in the training algorithm as well as the convergence time will vary depending on the weight initialization.

   Algorithm

- Step 1: First apply the inputs to the network and then calculate the output. The initial output can be anything as the initial weights were random numbers.
- Step 2: The error for neuron B can be calculated by Error B = output (1 − output B) (target B − utput B), where output (1 − output) is sigmoid function.
- Step 3: Now, if we want to change the weight, W + AB = WAB + (Error B × output A), where W + AB is new weight and WAB is the initial weight.
- Step 4: To calculate the errors of the hidden layer neurons. For hidden layer, we cannot calculate the error directly because we do not have a target value. By using back-propagation from output layer to hidden layer by taking the error from the output neurons, and then it is back-tracked through the weights to find hidden layer errors, i.e.,

Error A = output (1 − output A) (Error B WAB + Error C WAC)

---

[4]It is one of the most effective approaches to machine learning algorithm which has been developed by David Rumelhart and Robert McClelland (1994).

- Step 5: After getting error in the hidden layer neurons, then go to step 3 to change the hidden layer weights. Repeat these steps to train the neural network by using any number of layers.

   Hybridization of different algorithms has led to the creation of a trend known as soft computing. Based on that the NN[5] [1] algorithm has been considered it is to systemize the random search.

11.6 Architecture

See Fig. 11.2.

11.7 Validation Checks

Validation check #1:
The user Please select dataset for one party dataset in database. A message "Pls browse dataset" is displayed. Just check and leave.

Validation check #2:
The user Please select dataset for testing dataset in database. A message "Pls browse dataset" is displayed. Just check and leave.

11.8 Results and Description

See Fig. 11.3.
   Some diseases are listed as per the selection of symptoms the classification value is represented, which shows the symptoms of the patient. User or doctor will select

Fig. 11.2 Neural network algorithm diagrammatic representation

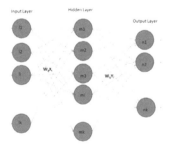

---

[5]Where the NN [1] has the ability to adapt to the circumstances and learn from the past experience.

Diagnosis of Chest Diseases Using Artificial Neural Networks

Network	Tuberculosis	COPD	Pneumonia	Asthma	Lung Cancer	Normal
GRNN	10	12	10	8	9	20
PNN	12	6	12	9	6	10
LVQ	20	10	13	12	10	9
RBF	11	12	8	23	5	19
LM	9	14	8	2	17	8
CFNN	2	2	4	5	6	7
FTDNN	5	23	6	8	21	9

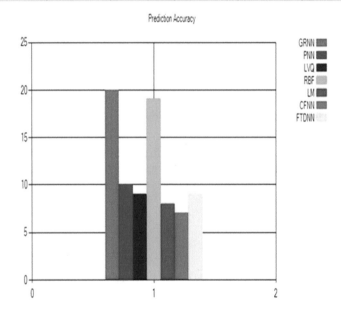

Fig. 11.3 Classification result of each neural network

the symptoms by selecting the checkbox. This selected checkbox is taken as input symptoms. As shown in a result 1 figure as represented above, the symptoms are selected and accordingly the diseases are predicted with classification values in the considered neural networks. In this way, the representation of some more results can be done to understand more clearly how the results are changed depending on symptoms.

11.9 Conclusion

In this study, the best result for the average classification accuracy was obtained using GRNN structure. The second best result for the classification accuracy was obtained using RBF structure. We have studied how the multi-layer system is used for medical diagnosis in chest disease. In this way, we are using the symptoms as an input which is been converted into value. These values and the extracted weights (randomly selected values) are used to calculate an activation function. The resultant output displays the type of disease using different types of neural network to predict the symptoms. Scope of this is to add the test report with the symptoms as inputs. Because of the different dataset used by the studies, the direct comparison of the results was impossible. So, these neural networks were compared using the same dataset which consists of the thirty-eight features.

References

1. M. Gori, A. Tesi, On the problem of local minima in back propagation. IEEE Trans. Pattern Anal. Mach. Intell. **14**, 76–85 (1992)
2. O.N. Yumusak, F. Temurtas, Chest diseases diagnosis using artificial neural network. Exp. Syst. Appl. **37**(2010)7648–7655
3. D.F. Speckt, Probabilistic neural networks. Neural Netw. **3**, 109–118 (1990)
4. A.M. dos Santos, B.B. Pereira, J.M. de Seixas, Neural networks: an application for predicting smear negative pulmonary tuberculosis, in *Proceedings of the statistics in the health sciences* (2004)
5. Application of neural network in diagnosing cancer disease using demographic data. Int. J. Comput. Appl. **1**(26) (2010)
6. Artificial neural networks in medical diagnosis. Int. J. Comput. Sci. Issues **8**(2) (2011)
7. K. Ashizawa, T. Ishida, H. MacMahon, C.J. Vyborny, S. Katsuragawa, K. Doi et al., Artificial neural networks in chest radiography: Application to the differential diagnosis of interstitial lung disease. Acad. Radiol. **11**(1), 29–37 (2005)
8. B.R. Celli, W. MacNee, Standards for the diagnosis and treatment of patients with COPD: A summary of the ATS/ERS position paper. Eur. Respir. J. **23**(6), 932–946 (2004)
9. D.F. Speckt, A generalized regression neural network. IEEE Trans. Neural Netw. 2(6), 568–576 (1991)
10. F.C. Chen, M.H. Lin, On the learning and convergence of the radial basis networks, in *Proceedings of the IEEE International Conference on Neural Networks*, (San Francisco, CA), pp. 983–98
11. G. Coppini, M. Miniati, M. Paterni, S. Monti, E.M. Ferdeghini, Computer aided diagnosis of emphysema in COPD patients: Neural-network-based analysis of lung shape in digital chest radiographs. Med. Eng. Phys. **29**, 76–86 (2007)
12. A.A. El-Solh, C.-B. Hsiao, S. Goodnough, J. Serghani, B.J.B. Grant, Predicting active pulmonary tuberculosis using an artificial neural network. Chest **116**, 968–973 (1999)
13. D.E. Rumelhart, G.E. Hinton, R.J. Williams, Learning internal representations by error propagation, in *Parallel distributed processing: Explorations in the microstructure of cognition*, vol. 1, ed. by D.E. Rumelhart, J.L. McClelland (MIT Press, Cambridge, MA, 1986), pp. 318–362
14. R.P. Brent, Fast training algorithms for multi-layer neural nets. IEEE Trans. Neural Netw. **2**, 346–354 (1991)

Chapter 12
Study on Efficient and Adaptive Reproducing Management in Hadoop Distributed File System

P. Satheesh, B. Srinivas, P. R. S. Naidu and B. Prasanth Kumar

Abstract The quantity of utilizations in view of Apache Hadoop is drastically expanding because of the robust elements of this framework. The Hadoop Distributed File System (HDFS) gives the unwavering quality and accessibility for calculation on applying static replication as a matter of course. Nonetheless, in perspective of the attributes of parallel operations on the application layer, the final result is absolutely unique for every information document in HDFS. Therefore, keeping up a similar replication instrument for each information record prompts to inconvenient consequences for the execution. By considering completely about the demerits of the HDFS replication, this paper initiated a methodology to deal with progressively obtaining the information document based on the predictive examination. With the assistance of likelihood hypothesis, the use of every information record can be anticipated to make a comparing replication procedure. In the end, the prevalent records can be thus reproduced by their own particular possibilities or by the low potential records, and an eradication code is connected to keep up the fixed quality. Thus, our approach all the while enhances the accessibility while keeping the dependability in correlation with the default method. Besides, the unpredictable decrease is connected to upgrade the viability of the expectation when managing big data.

Keywords Big data · Hadoop · Prediction model · Optimization

12.1 Introduction

The development of big data has made a fascinating change in application and arrangement advancement for separation, process, and load helpful information as it ascends to manage new difficulties. Around there, Apache Hadoop is one among the famous parallel systems. It is used not only to achieve high availability but moreover laid out to recognize and manage the disappointments and additionally keep up the information consistency. Joining the advancement of Apache Hadoop, the Hadoop Distributed File System (HDFS) was acquainted with the given dependability and high-throughput access for information-driven applications. Slowly, HDFS

© The Author(s), under exclusive license to Springer Nature Singapore Pte Ltd. 2019   121
P. V. Krishna et al., *Internet of Things and Personalized Healthcare Systems*,
SpringerBriefs in Forensic and Medical Bioinformatics,
https://doi.org/10.1007/978-981-13-0866-6_12

turned into an appropriate stockpiling system for parallel and circulated figuring, particularly for MapReduce motor, which was initially created by Google to adapt to the ordering issues on enormous information. To enhance the unwavering quality, HDFS is at first prepared with a system that consistently recreates three duplicates of each information record. This system is to keep up the prerequisites of adaptation to non-critical failure. Sensibly, keeping no less than three duplicates makes the information more dependable and more powerful while enduring the disappointments. Be that as it may, this default replication procedure still remains a basic downside as to the execution angle. Instinctively, the motivation behind developing Apache Hadoop was to accomplish better execution in data control as well [1]. In this way, this reason ought to be painstakingly learned at each part. In the execution point of view, in light of the notable research of postponement planning [2], when the errand set nearer to the data source is necessary, the framework accomplishes quicker calculation and efficient accessibility. The metric calculates the separation among the assignment, and the relating information source can be alluded to as the information region metric. The principle purpose behind the change is twice the number. In the first place, the system overhead can be decreased on runtime because of the accessibility of the nearby information; thus, no between correspondence is expected to exchange the required information from the remote nodes. Second, unmistakably the calculation of input information can begin instantly that is locally accessible; thus, no additional task scheduling exertion is expended. Subsequently, it is significant to state that enhancing the information territory would gigantically upgrade the framework execution regarding accessibility and estimation time. Potential is perceived as how as often as possible the particular record may be perused in whenever age. For instance, say a record has an accuracy of 32 inside the time of 5 s: This implies the record may be gotten to 32 times in the following 5 s. Moreover, the expected outcomes what's more, get to examples are stored in information base in request to right away match and rapidly condemn the appropriate activity without any need to recompute the comparative info. Once in a while, every information record can be proficiently imitated by an alternate in any case, suitable methodology. Furthermore, note that the final goal to keep up the adaptation to non-critical failure for less as often as possible got to information records, the open-source eradication code [3] is changed, connected to shield the framework against the impacts of disappointments. At long last, by actualizing this system, the undertaking execution time and capacity cost are enhanced profiting the profitability of huge information frameworks.

12.2 Related Work

Here, we discuss the three main basic concepts that are extremely important with respect to our work: cloud storage, replication management, and replica placement.

12.2.1 Distributed Storage

Accompanied by the fast development of Internet administrations, some extensive server frameworks were set up as server farms and distributed storage stages, for example, those in Google, Amazon, and Yippee!. Contrasted and customary substantial scale stockpiling frameworks worked for high-performance computing (HPC) concentrate on giving and distributing stockpiling administration on Internet which is touchy to application workloads and client practices. In addition, they give both adaptable information administration and a productive MapReduce [2] programming model for data-intensive processing. The key parts of distributed storage foundations are circulated document frameworks. Three popular illustrations are Google document framework [1] (GFS), Hadoop [3] appropriated record framework (HDFS), and Amazon Simple Storage Service (S3) [4]. Among them, HDFS is an open-source usage, so more subtle elements can be found. Working system of HDFS is like that of GFS; however, it is light-weighted. In GFS, three segments are customer, ace, and lump server, while in HDFS they are customer, name node, and data node as shown in Fig. 12.1.

A HDFS cluster comprises of a solitary name node, an ace server which deals with the document framework namespace and controls access to records by customers. Information nodes, typically one for each node in the cluster, oversee capacity appended to the nodes that they keep running on. Inside, a record is marked into at least one square, and these squares are put away in an arrangement of data nodes. The name node executes document framework namespace operations like opening,

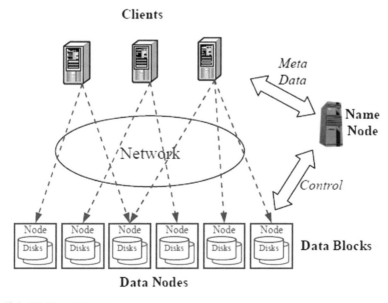

Fig. 12.1 HDFS architecture

shutting, and renaming documents and catalogs. It likewise decides the mapping of pieces into data nodes. The data nodes take charge of serving read and compose demands from the customers. The data nodes additionally accomplish square creation, erasure, what's more, replication upon direction from the name node. HDFS does not bolster simultaneous composes; just a single essayist is permitted at time.

12.2.2 Information Replication

The distributed storing framework includes documents that are typically divided into blocks over various information nodes to empower parallel gets to. In any case, striping diminishes document accessibility. Consider the likelihood of every information node accessible is p $(0 < p < 1)$, and record is divided into n $(n > 0)$ squares circulated as various information nodes. The entire record is accessible only if all the n squares having a place with this document are accessible. At that point, the likelihood of the record accessible is pn, and clearly, pn < p. The above investigation illustrates that accessibility debases because of stripping. Information replication has been broadly utilized as a mean of expanding the information accessibility of circulated stockpiling frameworks where disappointments are no more regarded like special cases. Copy number is the main issue of replication administration. Also to see how copy number impacts accessibility, execution, we took probes our model. Figure 12.2 demonstrates the relationship between accessibility and reproduction number when node disappointment proportion is 0.2 and 0.1. Considering the outcomes, we watch that accessibility enhances alongside the expansion of copy number. At the point when reproduction number meets a certain point, the document accessibility is equivalent to 1, and including more copies will not enhance the record accessibility anymore. To bring down the node disappointment proportion, it needs the low reproduction number.

Fig. 12.2 Availability varies with replica number

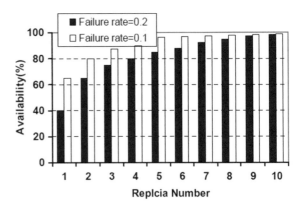

In this way, we can keep up least reproductions to guarantee a given accessibility necessity. With copy number expanding, the administration cost counting stockpiling and system data transfer capacity will altogether increase. Since a huge number of squares might be put away in every information node, even a little increment of reproduction number can bring about a critical increment of administration taken a toll in the general stockpiling framework. In a distributed storage framework, the system transfer speed asset is extremely constrained and significant to the general execution. An excessive amount of reproductions may not altogether enhance accessibility, but however bring pointless spending.

12.2.3 Replica Placement

An information node can at the same time bolster a predetermined number of sessions because of limit limitation. At the point where the quantity of periods has achieved its upper bound, association demands from application servers are blocked or dismissed. In expansive scale distributed storage framework, information nodes are heterogeneous, with various sorts of plates, system data transmission, CPU speed, and so on. Thus, maximal number of sessions that an information node can support is distinctive. Preventing likelihood in every information node might be distinctive because of various workload forces and limits [5]. Imitation situation impacts intra-ask for parallelism. Let us consider HDFS for instance. One favorable position of HDFS is that record can be exchanged among customers and information nodes in parallel. This may not be valid on the off chance that one of the intra-demand sessions is obstructed by information node. As shown in Fig. 12.3a, square B1 can be quickly served in the event that it has been repeated in data node2 rather than data node3. Copy arrangement likewise impacts between demand get to parallelism. Assume two customers ask for same square B1, as shown in Fig. 12.3b. Two solicitations can be instantly served if the data node1 has enough free sessions. Something else, the demand from client2 will be blocked and postponed. Regardless of the possibility that the piece B1 has been duplicated in the data node3, the demand still cannot be served promptly on the grounds that it does not have free sessions around then. In the event that imitation of piece B1 is set in the data node2 which has freely available sessions, the solicitations from client2 could be served instantly, increasing enhanced get to inertness. From the above examination, we can see that reproduction position has huge impact on blocking likelihood and get to skew. Productive copy arrangement can fundamentally help the inter-request, intra-demand parallelism, and general execution along with stack adjust of the HDFS cluster. This inspire us to initiate a novel imitation situation arrangement to effectively disperse work burden crosswise over cloud nodes, in this way enhancing blocking likelihood and get to skew.

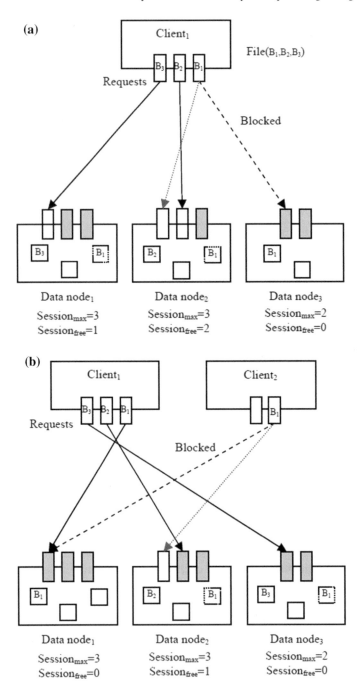

Fig. 12.3 a Replica placement affects intra-request parallelism. **b** Replica placement affects inter-request parallelism

12.3 Existing System

Various researches have been previously done on data replication in HDFS like achieving fault tolerance, [5–8] using data replication in HDFS. Fault tolerance is the main concept that was focused to overcome unexpected failures. In recent days, the research [9–11] is mainly concentrated over developing data locality for systematic execution on data replication in Hadoop and on some scheduled researches which have been initiated for improving data locality.

12.3.1 Data Locality Problem

Here, we discuss the data locality problem and its types in Hadoop. Data locality is related with the separation among information. If the distance between data and node is closer, then it is considered to have better data locality. Figure 12.1 describes the types of data locality in Hadoop: node locality, rack locality, and rack-off locality.

The data locality problem is defined as a status where a task is scheduled with rack or rack-off locality, and it might result in poor performance. Taking various aspects into consideration, the overhead of rack-off locality is greater compared to overhead of rack locality. To avoid this data locality problem, we put forth an efficient data replication scheme with the use of prediction by the access count of data files and a data replication placement algorithm which reduces rack and rack-off locality (Fig. 12.4).

12.4 Proposed System

12.4.1 System Description

The important task of the proposed model is to represent the replication factors and also dynamically and effectively form a schedule in order to place the replicas depending on the access potential of each data file. In addition to that, decrease the calculation time; the knowledge base and heuristic technique are developed to perceive the similarity in the access pattern among in-processing files and the predicted ones. As per the definition, the access pattern is really a group of eigenvectors mentioning the properties of processed information. Two files with similar access behaviors are regarded with a similar replication strategy. However, as a result these techniques are unit minor components and mainly employed in varied systems; discussing them is not inside the scope of this paper. Developed as a part of HDFS, the proposed approach (ARM) assumes liability in dealing with the replication over the HDFS nodes. Naturally, a review of ARM is portrayed in Fig. 12.1. As per given design, the traditional physical servers and also the cloud virtual machines can also

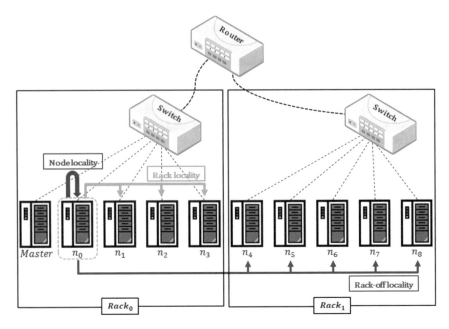

Fig. 12.4 Example of data locality

be utilized as and alluded to as nodes. ARM is appraised as a replication scheduler which can team up with any MapReduce work scheduler in the system configuration. Truth be told, ARM helps the FairScheduler and delay scheduling algorithm [2] to overcome demerits of larger tasks. Below is the description elucidating the operation of ARM (Fig. 12.5).

In the first place, the system begins by constantly gathering the heartbeat. From that point onward, this heartbeat is forwarded to the heuristic detector as the training information. This training information is contrasted with the access patterns, which are obtained from predictor component and stored at the database. On the off chance that there is a replica so that the access potential is recovered from the pattern and specifically passed to the predictor component with no calculation. If not, the training data is consistently moved rather as depicted in Fig. 12.2. In such case, the majority calculation falls under hyper-parameter learning and training phases of the prediction. In order to unravel such issue, the hyper-generator is being developed to diminish the computational complexity of the hyper-parameter learning stage. From that point onward, the training phase can begin to evaluate the access potential. At last, the access potential that belongs to target file is passed on to the replication management component. Additionally, for the next evaluation new pattern is obtained and stored at the database.

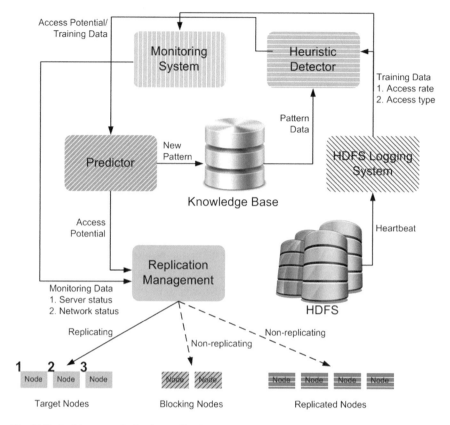

Fig. 12.5 Architecture of adaptive replication management (ARM) system

12.4.2 Replication Management

The motivation behind this segment is to depict how the replication management chooses the placement for the replica. Theoretically, by placing the potential replicas on low usage nodes (low blocking rate nodes), the replication management diverts the tasks to these idle nodes and balances the calculation. The blocking rate is computed and supported with the knowledge produced by the monitored system. Based on Ganglia framework, the monitoring system is simple, powerful, and simple to design for observing the vast majority of the required metrics. After plugging into the HDFS nodes, the monitoring system can gather statistics by means of Ganglia API. Since Ganglia gets a large portion of the metrics provided by HDFS, there is no contrast between this statistic and the heartbeat. Additional data is the only system statistic, which comprises of CPU system statistic, RAM usage, disk I/O, and network bandwidth. Such outline brings together the data sources for computational convenience, particularly to block rate calculation. So as to finish the replication management, we consider that the replication management component gathers all the constituents and produces the replication strategies. Based on this assumption, the access potential is used to scale the number of file copies. At that point, the main problem remaining is related to choosing the placement of the replicas. As described above, this duty is dependant on the statistics recovered from the monitoring system to ascertain the blocking rate and designate the replicas. Utilizing the parallel and distributed system theory, just a couple of complex factors can be taken into consideration to check the blocking rate of the server. These considerations incorporate the network bandwidth, the number of concurrent accesses, and the capacity of the server. The mechanism to figure the blocking rate is described below. With the help of considering the limited number of slots of node S_i as c_i, S_i may achieve a highly efficient blocking rate if the greater part among the slots has a tendency to be involved by the map tasks. It is obvious that the upgrading of data locality only profits the guide stage but not the shuffle and reduce phases. In this way, the blocking rate calculation is applied only to the map tasks. The probability of node S_i which is completely occupied by the map tasks defines the blocking rate of S_i, that is, denoted by $BR(S_i)$. Since the arrival rate T_i of the map tasks reaching the node S_i follows the Poisson distribution, the service process of S_i is carefully thought about to be of the $M = M = c$ queuing model. By definition, given $M = M = c$ Markov chain model is a stochastic process in which the first M stands for the Poisson arrival rate of customers, the second M denotes the exponential service rate of the servers, and c represents the capacity of each computing node (here, c represents the slot capacity). As a result, the blocking rate of S_i follows the Poisson arrival; see time averages (PASTA) theory as below:

$$BR(S_i) = \frac{(\lambda_i T_i)^{c_i}}{c_i!} \left[\sum_{k=0}^{c_i} \frac{(\lambda_i T_i)^k}{k!} \right]^{-1}$$

where T_i is the average mapping time of the foregoing tasks in S_i. Subsequently by estimating the blocking rate, it is easy for the replication management component to select a location to assign the replicas. As depicted in Fig. 12.1, the evaluated node fulfilling the two conditions is chosen as the destination. The primary condition is low blocking rate, and therefore, the other is not to store the required replicas beforehand.

12.5   Results

As a part of HDFS, the proposed system deals with replication over HDFS nodes. This mainly includes storing of desired replicas over HDFS nodes. On placing the potential replicas on low blocking rate, nodes balance the calculation thereby managing tasks. Also, proposed system depicts that high blocking rate is achieved through involving majority of the slots by map tasks. This makes simple for replication management component to choose an area to assign the replicas.

12.6 Conclusion

With a specific end goal to enhance the accessibility of HDFS by upgrading the information area, our commitment concentrates on taking after focuses. To start with, we plan the replication administration framework which is genuinely versatile to the normal information gathered for designing process. The approach not just effectively plays out the replication in the prescient way, additionally keep up the dependability by applying the eradication coding approach. Second, we propose a many-sided quality lessening technique to illuminate the execution issue of the forecast procedure. Truth be told, this many-sided quality lessening technique fundamentally quickens the forecast procedure of the get to potential estimation. At long last, we execute our technique on a genuine bunch and check the viability of the proposed system. With a thorough investigation over characteristics of the record operations in HDFS, our unique purpose is to make a versatile answer for progressing the Hadoop framework. For further improvement, a few sections of the source code are created to test our idea so as to make it accessible considering the terms of the GNU overall population permit (GPL).

References

1. C.L. Abad, Y. Lu, R.H. Campbell, Dare: adaptive data replication for efficient cluster scheduling, in Cluster (IEEE, 2011), pp. 159–168
2. Z. Cheng, Z. Luan, Y. Meng, Y. Xu, D. Qian, A. Roy, N. Zhang, G. Guan, ERMS: an elastic replication management system for HDFS, in *2012 IEEE International Conference on Cluster Computing Workshops (CLUSTER WORKSHOPS)*, Sept 2012, pp. 32–40
3. M. Sathiamoorthy, M. Asteris, D. Papailiopoulos, A.G. Dimakis, R. Vadali, S. Chen, D. Borthakur, Xoring elephants: novel erasure codes for big data, in *Proceedings of the VLDB Endowment*, vol. 6, no. 5 (VLDB Endowment, 2013), pp. 325–336
4. C. Guo, H. Wu, K. Tan, L. Shi, Y. Zhang, S. Lu, Dcell: a scalable and fault-tolerant network structure for data centers. ACM SIGCOMM Computer Communication Review **38**(4), 75–86 (2008)
5. S.B. Wicker, V.K. Bhargava, *Reed-Solomon Codes and Their Applications* (Wiley, 1999)
6. B. Calder, J. Wang, A. Ogus, N. Nilakantan, A. Skjolsvold, S. McKelvie, Y. Xu, S. Srivastav, J. Wu, H. Simitci et al., Windows azure storage: a highly available cloud storage service with strong consistency, in *Proceedings of the Twenty-Third ACM Symposium on Operating Systems Principles* (ACM, 2011), pp. 143–157
7. B. Fan, W. Tantisiriroj, L. Xiao, G. Gibson, *Diskreduce: replication as a prelude to erasure coding in data-intensive scalable computing* (Carnegie Mellon University, Pittsburgh, Parallel Data Laboratory, 2011)
8. C. Huang, M. Chen, J. Li, Pyramid codes: flexible schemes to trade space for access efficiency in reliable data storage systems. ACM Transactions on Storage (TOS) **9**(1), 3 (2013)
9. A. Datta, F. Oggier, Redundantly grouped cross-object coding for repairable storage, in *Proceedings of the Asia-Pacific Workshop on Systems* (ACM, 2012), p. 2

10. Y. Chen, S. Alspaugh, R. Katz, Interactive analytical processing in big data systems: a cross-industry study of map reduce workloads. Proceedings of the VLDB Endowment **5**(12), 1802–1813 (2012)
11. Z. Guo, G. Fox, M. Zhou, Investigation of data locality in map reduce, in *Proceedings of the 2012 12th IEEE/ACM International Symposium on Cluster.* Cloud and Grid Co

Printed in the United States
By Bookmasters